SOTHIS PRESS

Sothis Press aims to completely reimagine the public domain book with an emphasis on universal access to captivatingly designed print editions. We believe that technology will enable every single book ever published to be perpetually accessible in beautiful print form. Sothis is beauty in ubiquity.

www.sothis.press
1 Waterhouse Square, London, EC1N 2ST

Diagnostic Methods

A Guide for History Taking, Making of Routine, Physical Examinations and the Usual, Laboratory Tests Necessary for Students in Clinical Pathology, Hospital Internes and Practicing Physicians

Herbert Thomas Brooks

SOTHIS PRESS

ISBN 978-1-0368-0275-2

This edition is published by
Sothis Press © 2023. The company
is registered in England and Wales.
The Company Number is 14900389.

DIAGNOSTIC METHODS

A GUIDE FOR

HISTORY TAKING, MAKING OF ROUTINE PHYSICAL EXAMINATIONS AND THE USUAL LABORATORY TESTS NECESSARY FOR STUDENTS IN CLINICAL PATHOLOGY, HOSPITAL INTERNES, AND PRACTICING PHYSICIANS

BY

HERBERT THOMAS BROOKS, A.B., M.D.,

PROFESSOR OF PATHOLOGY, UNIVERSITY OF TENNESSEE, COLLEGE OF MEDICINE, MEMPHIS, TENNESSEE

THIRD EDITION

Revised and Rewritten

ST. LOUIS
C. V. MOSBY COMPANY
1918

Press of
C. V. Mosby Company
St. Louis

PREFACE TO THIRD EDITION

The fact of the necessity of a third edition, in so short a period of time following the second edition, indicates that this little book has been of some value.

As in the second edition, so also in this edition this book is intended for medical students, hospital internes, and physicians who have a limited amount of time, only, to give to laboratory work.

It is advised that along with this some extensive work on clinical diagnosis be used as a reference book.

The book has been carefully reviewed, all unnecessary or obsolete tests omitted, all new tests of any value have been added, and the whole matter gone over carefully in an effort to make everything clear, well defined, and to the point.

A new chapter on technique of staining and examination of smears, most important exudates, etc., has been added, which the author hopes will make it more serviceable to the practicing physician who is attempting to do his own laboratory work.

H. T. B.

Memphis, Tenn.

PREFACE TO SECOND EDITION

This little book is intended for medical students, hospital internes and physicians who have a limited amount of time to give to laboratory work.

Every possible care has been used to incorporate in this guide only the up-to-date and absolutely reliable laboratory tests. If one is able to perform these tests accurately and at the same time place the proper interpretation on the results, he will be very materially assisted in coming to an accurate diagnosis.

H. T. B.

Memphis, Tenn.
January, 1914.

CONTENTS

DIAGNOSTIC METHODS

CHAPTER I

OUTLINE FOR HISTORY TAKING

Obtain from the patient the following:

1. Name.
2. Age.
3. Occupation.
4. Social Relation.
5. Address.
6. Weight.
7. Height.
8. Color.

Chief Complaint and Duration of Symptoms

Secure from patient in as few words as possible the thing or things that give him the most trouble. Get date of beginning and duration of each.

Family History

If living, the health of, or if dead, the cause and age of death of father, mother, brothers, sisters, uncles and aunts. Secure positive or negative history of tuberculosis, cancer, rheumatism, gout, diabetes.

Personal History

Get the habits of the patient as to the use of alcohol, tobacco and drugs; habits as to sleep; exact nature of past and present occupations; condition of present surroundings, whether sanitary or not. If the patient is a woman, obtain the following:

Menstrual History.—Age of commencement; regularity; duration; quantity, whether copious or scanty, associated with pain (?); date of last menstruation; presence of leucorrhea.

Obstetrical History.—Number of children, age of youngest, number of miscarriages, date of last, character of labor, complications occurring during labor or immediately following.

Past History

Diseases of childhood, especially acute rheumatism, scarlatina, diphtheria, or tonsillitis. All other diseases or injuries prior to this illness. Ask your patient if he has had syphilis, or if he has been exposed to same. Also ask if he has had gonorrhea or a sore of any kind. Any acute infection, as typhoid or pneumonia?

History of the Present Illness

Chief complaint, date of beginning; onset, slow or sudden; order of appearance of symptoms, which gave patient the most trouble; character of treatment, if any.

Alimentary Canal—

If symptoms indicate disease of the alimentary canal, ask the following questions: (a) Appetite; (b) Meals—nature of the food patient most craves; (c) Sensations referred to the stomach—pain, fullness, or discomfort. Give location of pain and radiation, if any; (d) Vomiting—time of day, relation of food, relieves pain or not, any retching; (e) Character of vomited material—amount, color, red, coffee grounds, sour or frothy; (f) Eructations of gas; (g) Flatulence—relation to what particular kind of food, and does the gas tend to go up or down; (h) State of the bowels—as the presence of diarrhœa—is it associated with blood, slime or tenesmus, does any particular kind of food appear to bring it on? (i) Constipation—usual habits as to bowel movement; (j) Pain in the abdomen—where is it the worst, is it persistent or intermittent, is it relieved or made worse by pressure? Is there any radiation, any soreness following?

Liver—

If symptoms indicate the trouble to be in the liver, the following questions are to be asked: (a) Pain—any severe pain coming on and lasting a few hours only. Was the pain associated with vomiting? Does it radiate? Any pain in the shoulder? Was the patient yellow after the attack? (b) Piles; (c) Vomiting of blood; (d) Any change in the color of urine or feces; (e) Inquire as to the presence of digestive disturbances.

Cardiovascular System—

If the symptoms indicate that the trouble is in the cardiovascular system, the following questions should be asked: (a) Any dyspnœa; (b) Precordial pain or discomfort—does it radiate; (c) Any palpitation—relation to meals and exertion? (d) Headache; (e) Giddiness; (f) Sleep; (g) Any cough; (h) Presence of digestive disturbances; (i) Swelling of feet; (j) Presence of nose bleeding?

Respiratory System—

If the symptoms indicate that the trouble is in the respiratory system, the following questions should be asked: (a) Any cough—character of cough, when worse, any pain, associated with vomiting; (b) Expectoration—yellow or not, any blood present and how much, and whether only after severe coughing; (c) Any pain in chest—pain on deep breathing and where situated; (d) Dyspnœa—is it spasmodic, and describe the attack if it is; (e) Night sweats; (f) Loss of weight; (g) Loss of strength; (h) Fever; (i) Hoarseness.

Kidneys—

If the symptoms indicate that the trouble is in the kidneys, the following questions should be asked: (a) Any headache; (b) Drowsiness; (c) Dimness of vision; (d) Attacks of dyspnœa; (e) Vomiting; (f) Paralysis; (g) Convulsions; (h) Does the face look puffy in the morning; (i) Any swelling in the ankles; (j) Urine—altered in amount, clear or turbid, any blood, and is it present at the early or the latter part of urination; (k) Does the patient have to rise in the night to pass urine; (l) Is there any increase in frequency and is this by day or by night; (m) Pain on micturition? (n) Pain in lumbar region—does it radiate to the groin?

Nervous System—

If the symptoms indicate that the trouble is in the nervous system, the following questions may be asked: (a) Any headache (when worse, where located, character of the headache); (b) Nausea or vomiting; (c) Dizziness; (d) difficulty in walking; (e) Any transient palsies, or transient loss of memory; (f) Any visual disturbances; (g) Convulsions (general or local, age of beginning, frequency, any premonition, bite tongue, micturate or defecate); (h) Any discharge from ear at any time.

Blood—

If the symptoms indicate that the trouble is in the blood, the following questions should be asked: (a) Any dyspnœa? (b) Headache; (c) Dizziness; (d) Swelling of feet; (e) Previous attacks.

Bones and Joints—

If the symptoms indicate that the trouble is in the bones or the joints, the following questions may be asked: (a) Pain in bone—worse day or night; (b) Pain in joint—constantly present or only when the joint is moved; (c) Any sudden starting pains at night; (e) Is the pain affected by the weather; (f) Does the pain shift from one joint to the other?

Child—

If the patient is a child, the following questions should be put to the mother: (a) Number of children; (b) Any dead and the cause; (c) Any miscarriages; (d) Mother's health during pregnancy; (e) Is it a full time child; (f) Was the labor normal; (g) Was the child breast fed; (h) What food is being given the child at the present time; (i) Was there any rash or snuffles after birth; (j) State the age of cutting of teeth and walking; (k) Give the usual condition of the child's bowels; (l) Any bronchitis, measles, whooping cough, chicken pox, scarlatina, discharge from the ears.

CHAPTER II

PHYSICAL EXAMINATION OF THE PATIENT

General Survey of the Patient

Note the following: (a) Nutrition; (b) Attitude; characteristic of any trouble; (c) Expression, whether animated, apathetic or placid; (d) Complexion; (e) Eruptions; (f) Enlargement of lymph nodes or presence of any tumor masses; (g) Deformities, either congenital or acquired.

Examination of the Alimentary Canal and the Abdomen

Examine—

(a) LIPS.—Noting the presence of cyanosis, anæmia, herpes, or fissures; (b) TEETH.—Note the presence of caries, exposures of roots, pyorrhea, or congenital syphilis; (c) TONGUE.—Note the presence of tremor, any deviation on protrusion, size, papillæ, coating (moist or dry); (d) PALATE, FAUCES, and PHARYNX.—Note the presence of ulcers, mucous patches, enlarged tonsils, white patches on tonsils or uvula or soft palate; (e) ESOPHAGUS.—Any difficulty in swallowing, if necessary pass the stomach tube, and note presence of pain, stricture or diverticulum; (f) ABDOMEN.—*Inspection.*—Note general contour, the presence of pulsations in epigastrium, movements of abdominal wall, peristaltic waves, presence of striæ, pigmentations, or scars; *Palpation.*—Examine for the presence of tenderness, tumor, rigidity of muscles, and gurgling; *Percussion.*—Examine for dullness in the flanks, if present is it movable? Examine for presence of free gas in the peritoneal cavity; (g) STOMACH.—*Inspection.*—Examine for the presence of visible tumors, dilatations, and peristaltic waves; *Palpation.*—Examine for tenderness, tumors, and splashing; *Percussion.*—Examine for position of greater curvature, lower border of liver and lung. Inflate the stomach and examine for size and position.

LIVER.—*Inspection.*—Look for the edge of the liver and the presence of pulsations. *Palpation.*—Examine for tenderness, position of the lower border, regularity or irregularity of surface. *Percussion.*—Examine for upper border of dullness in the mid-clavicular line, mid-axillary line and scapula line.

SPLEEN.—Note whether or not palpable, position of the lower pole.

KIDNEYS.—Note whether or not palpable, presence of tenderness, and extent of mobility.

Examination of the Cardiovascular System

HEART.—*Inspection.*—Note the shape of the præcordial region, presence of pulsation—normal and abnormal in the præcordial region and immediately beyond, as in episternal notch, neck, to right of sternum, and epigastrium.

Palpation.—Location and description of apex beat, presence and location of thrills, and whether systolic, diastolic, or presystolic. Note the presence of a friction rub, and any abnormal pulsations.

Percussion.—Outline of superficial and deep cardiac dullness; description of any abnormal areas of dullness.

Auscultation.—Relative intensity of heart sounds in various valvular areas. Note the presence and give description of adventitious sounds, as cardiac murmurs, vascular murmurs, and pericardial friction rub.

PULSE.—Note the presence of arteriosclerosis; approximate blood pressure, whether low, moderate or high; character and quality of the pulse, especially whether characteristic of aortic insufficiency of stenosis. Note the presence or absence of venous pulsation in the neck.

Examination of the Respiratory System

THORAX.—*Inspection.*—Note the shape whether symmetrical or the presence of unilateral depressions or fullness. Note the presence of alar chest, flat chest, emphysematous chest, rachitic rosary, pigeon breast, Harrison's sulcus. Note intercostal spaces for undue fullness or retraction. Note the respiratory movements as to rate, type, equality or inequality of expansion. Note the presence of Litten's diaphragmatic sign.

Palpation.—Examine the respiratory movements as to equality or inequality of expansion. Note any localized deficiency of expansion. Note the presence and location of friction fremitus. Examine the vocal fremitus as compared with the normal.

Percussion.—Compare the corresponding areas of the chest. Outline the apices, lower borders of both lungs, and note whether the resonance is normal, diminished, or increased. Note the presence and location of any abnormal areas of dullness.

Auscultation.—Note the character of the breath sounds, whether normal, increased, diminished or absent. Note the presence and location of any abnormal areas of tubular, broncho-vesicular, cogwheel, or other abnormal types of breathing. Note the presence of any adventitious sounds, as rales (dry, bubbling, crepitant, or sub-crepitant). Also note pleuritic friction sounds. Note vocal resonance, whether normal, increased, diminished or absent.

Examination of the Nervous System

Cranial Nerves—

OLFACTORY NERVES.—Use oil of cloves, oil of peppermint or asafœtida. Compare the sense of smell in both nerves.

OPTIC NERVES.—Test the acuteness of vision, extent of visual fields in both eyes.

MOTOR OCULI, TROCHLEAR, and ABDUCENT NERVES.—Note the presence of strabismus, or ptosis. Test for defects in ocular movements, for diplopia, and for nystagmus. Note the size and shape of pupils. Test their reaction to light and accommodation.

TRIFACIAL NERVES.—Test the motor functions by having

the patient clinch the teeth tightly, and feel the contractions of masseter and temporal muscles. Test the sensory functions.

FACIAL NERVES.—Close the eyes tightly, wrinkle the forehead, and elevate the upper lip, noting any lack of symmetrical muscular action.

AUDITORY NERVES.—Test acuteness of hearing on both sides with watch. Examine for the presence of vertigo.

GLOSSOPHARYNGEAL NERVES.—Examine for taste on the posterior surface of the tongue by using a little sugar or quinine. Tickle the pharynx and note the presence of any gagging.

VAGUS NERVES.—Have patient to swallow liquid and note any regurgitation through the nose. Ask the patient to say "ah," and note if both the sides of the palate are raised equally.

SPINAL ACCESSORY NERVES.—Have the patient to shrug the shoulders, also to rotate the chin from side to side, noting any difference in the movements.

HYPOGLOSSAL NERVES.—Protrude the tongue, and note the presence of any deviation, or any atrophy.

Motor Functions—

Compare strength of corresponding groups of muscles in the upper and the lower extremities. Examine for muscular incoordination in the upper and lower extremities. Note state of nutrition of the muscles, and presence of any atrophy. Examine for abnormal muscular movements such as tonic spasms, clonic spasms, contractures, tetany, convulsions, intention tremor, fibrillary twitchings, choreic movements, athetosis. Examine for any muscular rigidity or flaccidity. Test for the presence of Kernig's sign.

Sensory Functions—

Test for common sensibility with feather or cotton. Note the presence of areas of anesthesia or hyperesthesia. Test for sense of pain. Use pin point and note the presence of areas of analgesia, hyperalgesia, or delayed conduction. Test for temperature sense, using test tubes with warm and cool water. Test for muscular sense, both as to sense of weight and position. Inquire as to any abnormal sensations, such as girdle pain, formications, pins and needles sensations, or numbness.

Reflexes—

SUPERFICIAL REFLEXES.—Test for: (a) Plantar reflex (Babinski's sign is an extensor plantar reflex occurring chiefly in pyramidal tract lesions). (b) Conjunctival reflex; (c) Pupil reflex; (d) Palate reflex; (e) Cremasteric reflex; (f) Abdominal reflex; (g) Epigastric reflex.

DEEP REFLEXES.—Test for: (a) Knee jerk; (b) Angle jerk; (c) Elbow jerk; (d) Jaw jerk; (e) Ankle clonus; (f) Patella clonus.

ORGANIC REFLEXES.—Inquire as to: (a) Deglutition; (b) Defæcation; (c) Micturition; (d) Incontinence, hesitancy or retention of urine.

Examination of the Locomotor System

Examine shafts of bones for signs of former fractures, thickening of periosteum, or the presence of tenderness. Examine the ends of the bones for enlargement as in rickets or nodules as in rheumatoid arthritis. Examine joints for the presence of swelling, tenderness, fluctuation, or redness. Observe the degree of motility in every direction. Examine the vertebral column for the presence of tenderness, local projections, lordosis, kyphosis, scoliosis, and mobility of the vertebræ.

EXAMINATION OF THE GAIT.—(a) Spastic gait, as occurs in hemiplegia; (b) Ataxia gait, as occurs in tabes; (c) Reeling gait, as in a drunkard; (d) Festinant gait, as in paralysis agitans; (e) Waddling gait as in pseudohypertrophic muscular paralysis; (f) High stepping gait, as in multiple peripheral neuritis.

ROMBERG'S SIGN.—Test for its presence.

Examination of Eyes, Ears, Nose, and Larynx

Eyes—

(a) Pupils (size, equality, shape, reflexes, Argyll Robertson); (b) Strabismus; (c) Ptosis; (d) Nystagmus (lateral or vertical); (e) Conjunctivitis; (f) Exophthalmos; (g) Vision (hemianopsia, condition of retina); (h) Oedema of lids.

Ears—

(a) Hearing; (b) Discharge; (c) Examination of canal for foreign bodies, or wax. (d) Examine tympanum; (e) Note any tenderness over mastoid.

Nose—

(a) Discharges; (b) Deformities; (c) Tumors; (d) Epistaxis; (e) Deviations of septum; (f) Presence of spurs; (g) Enlarged turbinates.

Larynx—

If there are any voice changes, a laryngoscopic examination is indicated.

CHAPTER III

SPUTUM

ORIGIN.—May be from mouth, nose, pharynx, larynx, bronchi or lungs. One or more or all. Note whether hawked up or coughed up.

QUANTITY.—The quantity varies within wide limits; as in early tuberculosis it is small, but in chronic bronchitis or bronchiectasis it is large.

ODOR.—Ordinarily there is no odor to sputum. In case of abscess or gangrene of the lung it may be very disagreeable.

SPUTUM FOR EXAMINATION should be coughed up and not hawked. In case of children it may be necessary to insert a swab in the pharynx, and as a result of this irritation the sputum will be coughed up and can be removed before it is swallowed.

WHEN A SPECIMEN OF SPUTUM IS RECEIVED, pour the sputum in a petri dish and place on cover. Place the sputum container in water and boil. Treat also in this manner the petri dish and sputum after examination.

Macroscopic Examination of the Sputum—

This shows the following:

1. *Mucous or Viscid Sputum.*—Seen in early stages of bronchitis and pneumonia.

2. *Mucopurulent Sputum.*—This is the most common form and is not characteristic of any particular condition.

3. *Purulent Sputum.*—This is seen in pure form only in perforation into the lungs or bronchi of foci of pus, as in abscess of lung or empyema.

4. *Serous Sputum.*—This is often slightly red in color and frothy. The red color is due to blood. This sputum is characteristic of pulmonary œdema.

5. *Nummular Sputum.*—Each part expectorated tends to collect to itself. This is common in tuberculosis of the lungs.

6. *Hemorrhagic Sputum.*—This is seen in phthisis, pneumonia, epistaxis, abscess of lung, hemorrhagic infarction, new growths and passive congestion.

7. *Tenacious Sputum.*—Adheres to an inverted cup. This is seen in pneumonia.

Note the Color of Sputum as follows:

1. Rusty or Orange Juice in color. This is common in pneumonia.

2. Prune Juice in appearance. Sometimes seen in pneumonia and often in cancer and gangrene of lungs.

3. Grass-Green. This is seen in pneumonia in combination with jaundice.

4. Black or Gray. This is due to substances inhaled, as coal dust. The sputum may be gray or black from food, as chocolate; and also tobacco.

5. Reddish-Yellow. Seen when abscess of liver ruptures into lung.

6. Hemorrhagic Sputum. Seen in pulmonary tuberculosis, pulmonary infarcts, passive congestion of lungs, lobar pneumonia, leaking aneurysm, and tuberculosis of lungs, etc.

Microscopic Examination Should Be Made for the following:

1. *Unimportant Constituents Often Seen.*—(a) leucocytes; (b) a few red blood cells; (c) epithelial cells, both squamous and columnar in type; (d) "heart failure" cells, pigmented epithelial cells from lining of the alveoli: results from passive congestion of lungs; (e) various bacteria; (f) particles of food.

2. *Important Constituents to Be Examined For:*

(a) Bacteria; this includes the tubercle bacillus, influenza bacillus, pneumococcus, streptococcus, staphylococcus and Friedländer's bacillus.

(b) Elastic Fibres; this is seen in all destructive processes of the lungs, as phthisis, gangrene and abscess.

(c) Curschmann's Spirals and Charcot-Leyden Crystals. For an examination, microscopically, of the sputum for leucocytes, epithelial cells, influenza bacillus, Friedländer's bacillus, streptococcus and staphylococcus, the Löffler's methylene-blue stain is sufficient.

This is as follows:

1. Make cover glass preparation from a purulent particle of the sputum and spread it thinly. Dry in air.
2. Fix by passing through flame three times.
3. Stain in Löffler's methylene-blue for 30 seconds, heating to the steaming point, or 1 to 2 minutes without heating.
4. Wash in water, dry, and mount in balsam.

Fig. 1.—Charcot-Leyden crystals.

Fig. 2.—Curschmann's spirals.

Method of Staining Bacillus Tuberculosis in Sputum—

1. Select a purulent or cheesy particle from the sputum and smear it thinly on a slide and let it dry in the air.
2. Fix the specimen by passing through flame four or five times.
3. Cover with carbol-fuchsin solution and steam from two to three minutes over Bunsen burner or alcohol lamp. Do not let the stain dry, but add more of the stain if necessary.
4. Wash in water.
5. Decolorize in acid alcohol until a pinkish tinge remains.
6. Wash in water.
7. Cover the specimen with Löffler's methylene-blue solution for 30 seconds.
8. Wash in water, dry, mount in balsam and examine with oil immersion lens. The tubercle bacilli are stained bright red, nuclei and other bacteria are blue.

Antiformin Method for Demonstrating Tubercle Bacilli—

To 5 to 10 c.c. of sputum add 5 to 10 c.c. of a 30 per cent solution of antiformin. Stir for a short time with a glass rod to hasten fluidification. Let stand for one or two hours, centrifugalize for three or four minutes. Pour off supernatant fluid, add physiological salt solution and mix with a glass rod, centrifugalize again, pour off the supernatant fluid, add more physiological salt solution and mix again. Repeat all this one more time. Then transfer sediment to glass slide, smear over the surface, dry and fix in the usual way. Stain for tubercle bacilli as described above.

Method of Staining the Pneumococcus—

The pneumococcus may be stained by the Löffler's methylene-blue solution, but probably the most satisfactory method is Gram's stain.

Reagents required for Gram's stain:

1. Anilene water, made by placing 3 to 5 c.c. of anilene oil in test tube, and four times this quantity of tap water. Shake well, and filter through two sheets of filter paper.
2. Anilene water gentian violet, made by adding two or three drops of saturated alcoholic solution of gentian violet to the anilene water, and filter.
3. Gram's iodine solution. (See composition of reagents.)
4. Dilute carbol-fuchsin: Made by adding two (2) parts of water to one part of carbol-fuchsin, then filter.

Following is the Method of Gram's Stain:

1. Cover the cover glass containing the smear with anilene water gentian violet for 2 to 3 minutes; blot dry.
2. Gram's iodine for 1½ minutes; blot dry.
3. 95 per cent alcohol; pour on and off until all the blue color comes away.
4. Wash in water.
5. Dilute carbol-fuchsin (1 to 2 with water) for one-half to one minute.

Wash in water, mount and examine. The pneumococcus and all Gram positive organisms are colored blue. All Gram negative organisms are colored red.

Examination for Elastic Fibers—

Take 8 c.c. of sputum and equal amount of 10 per cent sodium hydrate, shake well and heat the mixture (but keep below the boiling point) until whole mass is liquified. Dilute with water, centrifugalize. Pour off supernatant fluid. Make a thick smear on slide from the sediment and let dry in the air. Cover smear with Weigert elastic tissue stain and steam for 4 minutes, but be careful to prevent alcohol from catching fire. Wash gently in water. Decolorize for one minute with 95 per cent alcohol. Wash in water. Dry and mount and examine under dry lense for elastic fibres, which appear blue and wavy.

An examination should be made for *Curschmann's Spirals* and *Charcot-Leyden Crystals* in the sputum of every case of bronchial asthma. Use a two-thirds objective. The Curschmann Spirals appear as twisted spirals of glossy transparency. Macroscopically they appear as twisted shreds about 1 m.m. thick and 1 to 2 centimeters long, and are composed of mucous threads. The Charcot-Leyden crystals are not so frequently present. They are transparent and octahedral crystal.

Sputum in Disease

Tuberculosis—

In the *early stages*, none at all or small in amount; occurs most often early in the morning, and has a mucous or slight mucopurulent appearance. This may contain tubercle bacilli. As the disease progresses it becomes more and more mucopurulent, and finally purulent with cheesy-looking particles. It is then nummular in character. Amount of sputum gradually increases from almost none in the early stages to large quantities in the advanced stages.

Blood is present in almost all cases at some stages of the disease. In early cases the sputum is only streaked with blood, while larger hemorrhages are more apt to be seen later. Cheesy particles are observed in the moderately advanced and advanced cases. They usually contain tubercle bacilli in large numbers.

Sputum in Pneumonia—

In young children and the very old often there is no expectoration. The sputum is at first mucoid, but sooner or later becomes bloody to a greater or less degree, and in fully one-third of the cases it is rusty in appearance. It is very tenacious, and adheres to the cup when inverted. Prune juice sputum is often but not always an evil omen, as it indicates a serious condition.

In cases of abscess or gangrene, complicating the pneumonia, the sputum is more fluid; disagreeable odor especially in gangrene, and the color is coffee-like or chocolate brown. Microscopically one finds pus cells, epithelial cells, red blood cells and pneumococci.

Sputum in Bronchiectasis—

Usually very abundant, and often mouthful at a time; color is grayish-yellow, which may show red or brown pigment; odor usually disagreeable, especially if putrefactive changes have taken place in the bronchitic cavities.

Microscopically one finds pus cells, alveolar epithelial cells, few red blood cells, and an immense number of bacteria.

Sputum in Bronchitis—

Early in the attack the sputum is scanty, mucoid, and highly tenacious, occasionally streaked with blood. As the bronchitis continues the sputum becomes progressively more purulent, which gives it a yellow color.

Microscopically one finds leucocytes, ciliated epithelial cells, a few blood cells, and bacteria.

Sputum of Bronchial Asthma—

During the paroxysm there is often no sputum. If present it is scanty in amount, and occurs as grayish mucoid masses. As the attack passes off sputum appears. It is then fairly abundant, thin and frothy and contains mucopurulent masses. The sputum contains Curschmann's Spirals, mucous moulds of the smaller bronchi, and sometimes fibrinous casts.

Microscopically one finds many eosinophilic leucocytes, which is of some diagnostic importance. Charcot-Leyden crystals are often found.

CHAPTER IV

URINE

Get quantity in 24 hours. In suspected nephritis separate day and night portions. Note color, odor, reaction, and specific gravity.

Specific Gravity

As a rule, this is of value only in a 24-hour specimen. Place urine in a glass cylinder carefully so as not to give rise to formation of a foam. Should foam collect, remove with filter paper. Use a sufficiently wide cylinder to prevent the urinometer touching the sides. Insert urinometer gently and make your readings with your eye on a level with the bottom of the meniscus (concave upper surface of the fluid). Ordinarily the instruments are standardized for use at a temperature of 15° C. For each 3° C. above this temperature, the specific gravity is changed one point—in the third decimal point. As an example, if the specific gravity of a urine were found to be 1.016 at 28° C., the corrected reading would be 1.016+.004=1.020. Ordinarily the specific gravity of a 24-hour specimen ranges between 1.016 and 1.022, but 1.012 to 1.028 would still be in the limits of the normal.

Preservation of Urine

Specimens may be preserved by:

1. The addition of 40 per cent formalin in the proportion of 30 drops to a liter of urine.
2. Cold storage.
3. Chloroform in the proportion of 20 drops to 4 ounces. Use tightly corked bottle.
4. Thymol, 2 or 3 crystals to 4 ounces.
5. Toluol. Cover the urine with one-eighth of an inch of toluol. It is of special value in the preservation of volatile substances, as, diacetic acid.

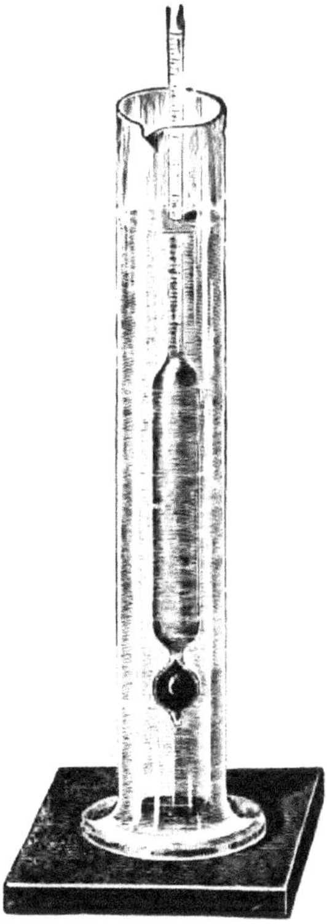

Fig. 3.—Urinometer.

Turbidity

Cloudiness of urine may be due to the precipitation of phosphates, or urates, or may be due to pus, blood, fat, or bacteria. If cloudiness is due to urates it will clear up on heating; if due to phosphates, with a few drops of acetic acid; if due to fats it will clear up by shaking with equal quantity of ether. If due to bacteria. clear up by the addition of aqueous solution of ferric chloride and then ammonium hydrate, and then filter. Cloudiness from red blood cells, pus, bacteria and cellular elements may be determined by microscopical examination.

Chemical Examination of the Urine

A qualitative examination should be made for the following abnormal constituents of the urine: (a) Any albumen; (b) Serum albumen; (c) Albumose; (d) Bence-Jones albumen; (e) Nucleo-albumen; (f) Serum-globulin; (g) Hæmoglobin; (h) Bile; (i) Indican; (j) Sugar (k) Acetone; (l) Diacetic acid; (m) Beta-oxybutyric acid; (n) Diazo substances; (o) Urobilin.

Chemical Test for Albumen—

1. *Nitric Acid or Heller's Test.*—This reacts to all urinary proteids except peptone. Place 5 c.c. or thereabouts of colorless nitric acid in the bottom of a test tube. Incline the test tube, and place in equal amount of clear filtered urine with a pipette. A white precipitate of albumen forms at the

junction of the two fluids, if albumen is present in the urine. A cloudy precipitate higher in the urine may be due to urates. Bile and an excess of urinary coloring matter may give a colored precipitate at the junction of the two fluids. This is a very accurate test, as it shows fairly definitely even to as small amount as 1/1000 of 1 per cent of albumen in the urine.

2. *Heat Test.*—Pour 10 c.c. of clear filtered urine into a test tube, and 1/10 of its volume of saturated sodium chloride solution is added if specific gravity is less than 1012, and then boil the upper half of the fluid. Add 3 to 4 drops of 25 per cent acetic acid and boil again. A precipitate appearing on boiling, which persists after the addition of the acid, or appearing on the second boiling, is albumen. One disappearing with the addition of the acid is phosphates. The test may fail if too much acetic acid is added. This is one of the most delicate tests for albumen.

Acetic Acid and Potassium Ferrocyanide Test—

Make 8 or 10 c.c. of clear filtered urine strongly acid with 25 per cent acetic acid—12 to 15 drops. Then add drop by drop a 5 per cent solution of potassium ferrocyanide. A white flocculent precipitate is formed if albumen is present.

Picric Acid Test—

To 8 or 10 c.c. of clear filtered urine add a few drops of Esbach's reagent. The appearance of a precipitate indicates presence of albumen.

Chemical Tests for Serum-albumen—

Add to a test tube half filled with filtered urine one-fifth of its volume of a saturated aqueous solution of sodium chloride, heat to a boiling point; add 2 to 5 drops of 50 per cent acetic acid and heat again. The persistence of a precipitate indicates serum-albumen.

Chemical Test for Nucleo-albumen—

Place in a test tube 5 c.c. of urine, and nearly fill test tube with water. Divide this into equal parts. Use one as a control, and to the other add 4 drops of 50 per cent acetic acid without heating. A cloudiness indicates the presence of nucleo-albumen.

Chemical Test for Serum-globulin—

Fill two test tubes about two-thirds with distilled water. Use one as a control, and the other add 10 drops of urine drop by drop. A cloudiness indicates the presence of serum-globulin.

Chemical Test for Albumose—

Take 8 c.c. of urine and add nitric acid, drop by drop, till you get a permanent precipitate. Heat to the boiling point and the precipitate will partially disappear. Cool upper part with water and it will reappear. Repeat this. The presence of albumose is indicated by the occurrence of a precipitate, which disappears on boiling and reappears on cooling. The nitric acid converts the albumen into an acid albuminate which is soluble, but the albumose is not affected.

Mix 10 c.c. urine with saturated solution of sodium chloride. Add 1½ c.c. of a 40 per cent sodium hydrate solution and shake tube thoroughly. Heat upper part of tube to boiling and add 10 per cent lead acetate solution drop by drop, continuing the heating. A brown to black precipitate indicates this form of albumen.

Chemical Examination for Hæmoglobin—

To about 10 c.c. of urine in a test tube add 2 c.c. of glacial acetic acid and 10 c.c. of ether. Shake gently for 1 or 2 minutes. After the ether has separated, decant. Add to the ethereal solution 10 drops of freshly prepared tincture of guaiac and 30 drops of hydrogen peroxide. A blue color indicates the presence of blood or blood coloring matter (hæmoglobin).

Chemical Examination for Bile—

Iodine Test. (Tinc. iodine 1 part, alcohol 15 parts.) Pour 2 c.c. of the iodine solution on the top of the urine in a test tube. A green ring at the junction of the two fluids shows bile.

Gmelin's Test for Bile—

The urine is treated with concentrated nitric acid as for Heller's albumen test, when in the presence of bile, a play of colors appear, green, blue, violet, and reddish-brown, the green being the most characteristic.

Chemical Examination for Indican—

(1) To 8 or 10 c.c. of clear filtered urine add about 2 c.c. of 5 per cent copper sulphate solution, and 3 c.c. of chloroform, and a quantity of concentrated hydrochloric acid equal to number of c.c. already present. Invert test tube a few times gently. If indican is present in excess a bluish color in the chloroform as it settles to the bottom of the tube.

(2) To about 10 c.c. of clear filtered urine add equal quantity of concentrated hydrochloride acid which contains in 100 c.c. .4 gram of ferric chloride. Shake well, then add 3 to 4 c.c. of chloroform and shake again. The chloroform will become blue if indican is present.

Sugar—

If albumen is present in more than a trace, the urine should be acidified with several drops of 25 per cent acetic acid, heated to precipitate the albumen and then filtered.

1. *Fehling's Test.*—Add equal quantities of the copper solution and alkaline solution. Dilute 3 to 4 times with water. Boil the upper fourth of the Fehling's solution and then add, at once, 1 or 2 drops or more of the urine. If sugar is present you will get a reddish or yellowish precipitate. Do not boil a second time if there is no precipitate, but set aside for a few hours and then examine for precipitate. If you should boil a second time and then no reddish precipitate appears there is no sugar present, but a reddish precipitate might then be due to other things in the urine than sugar.

2. *Phenyl-hydrazine Test.*—To 5 c.c. of the clear filtered urine add 5 c.c. of the phenyl-hydrazine acetate solution. Heat to the boiling point in the water bath for 30 to 45 minutes, then allow it to cool gradually. Place some of the yel-

low precipitate on a slide and then examine microscopically for the needle-like crystals of phenyl-glucosazone. Both glucose and levulose give identical crystals.

3. *Nylander's Test.*—Add 1 c.c. of Nylander's reagent to 10 c.c. of clear filtered urine in a test tube. Boil upper part 2 or 3 minutes. If sugar is present the fluid assumes first a yellow, then yellowish-brown, and finally an almost black color, and after some time a black sediment forms.

4. *Fermentation Test.*—To some of the urine add a little fresh yeast. Fill the fermentation tube provided for the purpose with the mixture. In a second fermentation tube place water containing a little yeast. In a third fermentation tube place some of the urine without yeast. Place in a warm room, or in an incubator until the next day, and note the presence of the CO_2 gas. This test proves that the reducing substance is a fermentable sugar, and in the vast majority of instances the sugar is glucose.

5. *Trommer's Test.*—Add to 5 c.c. of the clear filtered urine an equal volume of strong sodium hydrate solution, then drop by drop a 1 per cent solution of copper sulphate as long as the precipitate formed continues to dissolve readily. Heat upper part, and a reddish-yellow precipitate means the presence of glucose.

Acetone—

This substance may occur in the urine under many conditions, but its recognition is of value only in diabetes mellitus. Here it is practically always present and the amount is proportionate to the severity of the case, increasing as the patient becomes worse.

Lieben's Test is probably the simplest. To 5 to 10 c.c. of the urine add enough strong caustic soda solution to render the urine strongly alkaline and add about 1 c.c. of a dilute Lugol's solution. If any acetone be present, a precipitation of iodoform crystals occurs. This may be recognized by its odor or heating or by the form of the crystals under the microscope.

Diacetic Acid—

The presence of diacetic acid in the urine of diabetes is of even more importance than acetone, as it is usually the forerunner of diabetic coma.

Gerhart's Test is the one most commonly used. To a few c.c. of urine in a test tube add a strong aqueous solution of ferric chloride drop by drop. This precipitates the phosphates. When all the phosphates have been thrown down, filter the urine. If the resulting filtrate is a Bordeaux red color it may mean the presence of diacetic acid. In order to prove it, take another portion of the urine, boil it thoroughly and treat it in the same manner. As diacetic acid is decomposed by heat and volatilized as acetone, the Bordeaux red color will be absent or less intense in the boiled specimen if diacetic acid were present in the original, while if the red color were due to some other substance, there will be no change in the boiled specimen.

Beta-oxybutyric Acid—

In testing for diacetic acid, if the ferric chloride reaction is strongly positive Beta-oxybutyric acid is probably present.

Diazo Substances—

To 5 c.c. of sulphanilic acid solution add 2 drops of a 5 per cent fresh solution of sodium nitrite. Add an equal quantity of urine. Shake and add quickly 2 or 3 c.c. of 10 per cent ammonium hydrate solution. A carmine color, especially in the foam, shows a positive diazo reaction. If the reaction is positive and the mixture is allowed to stand for 24 hours a precipitate forms, the upper margin of which exhibits a green, greenish-black or violet zone.

Urobilin Test—

To 8 c.c. of unfiltered urine add an equal amount of a saturated solution of zinc acetate in absolute alcohol. Shake, add a few drops of Lugol's solution and filter. A fluorescence in the urine shows the presence of urobilin. The intensity of the fluorescence indicates the extent of the blood destruction.

This appears in the urine when there is much destruction of red blood cells, as in pernicious anemia and malaria cachexia.

Quantitative Examination of the Following Normal Constituents of the Urine

(a) Chlorides; (b) Urea; (c) Total acidity; (e) Total solids.

Chlorides—

Take 10 c.c. of clear, filtered urine; add to this 50 c.c. of distilled water, 5 c.c. of 5 per cent ammonium iron alum solution, 10 c.c. of concentrated nitric acid, and 20 c.c. of standard silver nitrate solution. Increase the quantity to exactly 100 c.c. with distilled water, shake and filter; measure out 50 c.c. of the filtrate in a beaker, and titrate with standard ammonium sulphocyanate solution until a reddish-brown tint first extends throughout the whole liquid.

The amount of sulphocyanate used multiplied by two gives at once the excess of silver nitrate used beyond the quantity required to precipitate all the chlorides in the 10 c.c. of urine. Knowing the original number of c.c. of silver nitrate taken and the excess, the difference will give the number used up, and each c.c. of this equals 10 mgms. of sodium chloride. This is multiplied by (1/10) one-tenth of the 24-hour quantity of urine to get amount of sodium chloride excreted in 24 hours.

Urea—

Use Doremus-Hinds ureameter. Fill the large tube with a mixture of sodium hydrate solution and bromine solution in the proportion of 15 c.c. of the former to 1 c.c. of the latter. Fill the smaller tube to the zero point with urine. Now very gradually allow 1 c.c. of the urine to pass into the mixture in the large tube. Urea is decomposed and nitrogen gas is liberated and collects at the top. If the level of the fluid stands at .015, then 1 c.c. of the urine contained .015 grams of urea. This must be multiplied by number of c.c. in 24-

hour quantity to get total quantity of urea passed in that length of time.

The Doremus ureameter differs from the Doremus-Hinds in that a 1 c.c. graduated pipette is non-attached and is therefore less convenient. One c.c. of the urine is added to the solution and, after the bubbles of gas have ceased to rise, make the reading and estimation in the same manner as the above.

Total Acidity—

Take 25 c.c. of urine, add to this 2 c.c. of saturated potassium oxalate solution; then two or three drops of 1 per cent phenolphthalin solution. Titrate this with decinormal sodium hydrate solution until the first change to red color takes place. This represents the neutralization of all the acids in 25 c.c. of the urine. From this estimate quantity in a 24-hour specimen. The normal is about 400 expressed in terms of decinormal sodium hydrate solution.

Total Solids—

This is estimated by multiplying the last two figures of the specific gravity of a mixed 24-hour urine by 2.33. Then multiply the product by the number of cubic centimeters voided in 24 hours and divide by 1,000. This will give approximately the total solids expressed in grams.

Quantitative Examination of the Following Abnormal Constituents of the Urine

(a) Albumen; (b) Sugar; (c) Indican.

Albumen—

An approximate idea of the quantity of albumen can be obtained by boiling 10 c.c. of urine acidified with two drops of 50 per cent acetic acid and allowing the albuminous precipitate to settle for 24 hours. If the albumen amounts to 2 or 3 per cent the fluid will be converted into an almost compact coagulum. One per cent of albumen in the urine the precipitate will occupy half the column of urine, five-tenths per cent of albumen will occupy one-third, one-tenth per cent will occupy one-tenth the volume, five-hundredths per cent will cover the bottom of the test tube. One-hundredth per cent or less causes a turbidity, but no precipitate.

Method of Esbach.—This method, while not accurate, is convenient and applicable. Fill the Esbach albuminometer to the line marked "U" with the urine, and fill to the line marked "R" with Esbach's reagent. Close the tube, and by repeated inversions thoroughly mix the two fluids. Do not shake. Set aside in natural position for 24 hours, when the precipitated proteid will settle to the bottom. The graduations indicate the grams of proteids in a liter of the urine. If amount of proteid is large, it will be necessary to dilute urine.

Fig. 4.—Esbach albuminometer.

Tsuchiya's Modification of Esbach's Method.—Acidify urine with a few drops of 25 per cent acetic acid. Fill Esbach tube with urine to mark "U," and then Tsuchiya's reagent to mark "R." Cork and invert 12 times to mix well. Place in vertical position at room temperature for 24 hours, and then read as in Esbach's method. This test is excellent in many respects.

Indican—

To 10 c.c. of urine add 1 c.c. of 25 per cent lead acetate solution. Filter. To 5 c.c. of the filtrate add 5 c.c. of Obermayer's reagent (concentrated hydrochloric acid, 500 c.c. ferric chloride, 1 gm.). Shake gently. Add 2 c.c. of chloroform. Shake and let stand for 5 minutes. A blue color indicates the presence of indican. Now drop from a burette a solution of potassium chlorate (34.64 gm. to a liter of water) until the blue color disappears. It normally requires about 3 drops. If indicanuria is present it may require 10 to 15 drops or more.

Sugar: Fehling's Quantitative Method—

Solution A.—Copper sulphate solution. Made by weighing out accurately 34.64 gms. of purest obtainable copper sulphate, and dissolve in exactly 500 c.c. of distilled water. 5 c.c. equals .05 gm. of glucose.

Solution B.—Alkaline tartrate solution. Dissolve 125 gms. of sodium hydroxide, and 173 gms. of Rochelle salts in 500 c.c. of distilled water. Place 5 c.c. of the copper sulphate in a moderately large flask, and add an equal volume of the alkaline tartrate solution. Dilute this with 5 or 6 volumes of water, and bring to a boil. Fill a burette with undiluted urine to be tested, and run it into the flask until the blue color of the solution has entirely disappeared. In the beginning of this titration .5 c.c. of urine may be added at a time with boiling after each addition. Towards the end, when the blue color has almost disappeared, the urine should be added more cautiously, 2 or 3 drops at a time. In order to be accurate, the urine for the final determination should be so diluted that not less than 5 c.c. would be used in the titration. Therefore, three determinations should be made, the first with undiluted urine to obtain an approximate estimate only and to find out the extent of the dilution in order that at least 5 c.c. of the urine will be used in the titration. The last two determinations to be made with the diluted urine, the average being taken.

The amount of copper sulphate solution used is reduced by .05 gm. of glucose. This amount of sugar, then, must therefore be in the volume used for the titration. Calculate the per cent.

Sugar: (Quantitatively)—Modified by Rudische—

To 4 parts by volume of a 50 per cent solution of potassium sulphocyanate, chemically pure, is added 1 part by volume of a mixture of equal parts of Fehling's copper sulphate and alkaline solutions. Place 25 c.c. of this solution in a beaker, and the urine to be tested added drop by drop from a burette until the blue color of the copper entirely disappears. Throughout the titration the solution should be slowly boiled and constantly stirred with a glass rod. The end reaction is sharp, the fluid becoming colorless or assuming a faint yellow tinge. Each c.c. of the reagent is reduced by 1 M. gm. of sugar: therefore it takes 25 M. gms. of sugar for the 25 c.c. of the reagent, and this amount of sugar is present in the quantity of urine used in the titration.

Benedict's Method for Quantitative Determination of Glucose—

Three solutions necessary:

Solution A.—Crystallized copper sulphate 69.3 gm., distilled water to 1,000 c.c.

Solution B.—Pure Rochelle salt 346 gm., anhydrous sodium carbonate 200 gm., distilled water to 1,000 c.c.

Solution C.—Potassium sulphocyanate 200 gm., distilled water to 1,000 c.c.

Take 10 c.c. of each of the three solutions; this gives together 30 c.c. This is equivalent to .073 gm. glucose. Place the 30 c.c. mixture of A, B, and C solutions in an Erlenmeyer flask, add 5 gms. of crystallized sodium carbonate and bring the mixture to a boil. When the sodium carbonate is in solution, allow urine to run from a burette until the last trace of blue color disappears. The end reaction is sharp. The amount of urine used in titration contains .073 gms. of glucose. Calculate the amount in 100 c.c. to obtain the percentage of glucose. Before titration the urine is preferably treated in the following way: Shake 50 c.c. of urine with 3 gms. of basic lead acetate and filter. Treat filtrate with neutral potassium oxalate, stir and then filter. Fill burette with this filtered urine and titrate as described above.

Examination of the Urine for Bacillus Tuberculosis

1. Centrifugalize, decant, dilute with water and recentrifugalize. Make a cover glass preparation from the sediment. This should be spread thinly and dried by holding over the flame of a Bunsen burner. Fix by passing through the flame three times.

2. Cover the preparation with carbol-fuchsin and steam over a flame for 2 minutes. Do not let slide become dry, but add more stain if necessary.

3. Wash in water.

4. Decolorize in acid alcohol until the red color is removed and faint pink is seen.

5. Wash in water.

6. Cover with Löffler's methylene-blue for 30 seconds.

7. Wash in water and mount.

The tubercle bacilli are bright red; nuclei and other bacteria are blue. The alcohol decolorizes any smegma bacilli, but absolute differentiation requires inoculation of guinea-pig.

Microscopic Examination of the Urinary Sediment

Centrifugalize with hand, water or electric centrifuge, and examine for crystals, fat globules, epithelial cells, red blood cells, pus cells and casts. Use two-thirds and one-sixth lens. Urine should be fresh, as casts, etc., easily disintegrate, especially in warm weather, or in a warm room.

Epithelial cells are recognized by the presence of a nucleus in the center of the cell. The cells may be of various sizes and shapes, depending on whether they originate from urethra, bladder, ureter or kidney.

Red cells are recognized by their spherical shape, and absence of nucleus. They may appear only as circular rings or may contain some hæmoglobin.

Pus cells (leucocytes) appear spherical and granular. On addition of a drop of dilute acetic acid the nucleus will become visible and its shape may be made out.

Casts are recognized by the following characteristics: Their sides are parallel, their ends are square or somewhat irregular, but they never come to a point. They are divided into the following:

1. Hyaline casts; (a) narrow, (b) broad.

2. Granular casts; (a) finely granular, (b) coarsely granular.

3. Waxy casts.

4. Fatty casts.

5. Casts containing organized structures; (a) epithelial casts; (b) blood casts; (c) pus casts.

Casts must be distinguished from mucous threads and cylindroids. Mucous threads show faint longitudinal striations and taper to a point at both ends. Cylindroids are homogenous like hyaline casts, but one or the other end tapers to a point.

Urine in Disease

Acute Diffuse Nephritis—

Quantity of urine very much diminished, depending of course on its severity.

Specific gravity and color are correspondingly increased, the color is often smoky from the presence of blood. In other cases the microscope is required to demonstrate the blood. Albumen is present in large amounts, 1/8 to 1 per cent.

Microscopical examination shows the presence of hyaline, granular and epithelial, and usually also blood and leucocytic casts. In addition there are renal epithelial cells which sometimes show evidence of fatty degeneration. Leucocytes and red blood cells in varying numbers are seen.

Chronic Diffuse Nephritis (Large White Kidney)—

Urine diminished in quantity. Specific gravity and color correspondingly increased. Albumen is always present and usually in a greater amount than in the acute cases; the amount varies from 1/4 to 1 per cent.

Microscopically one finds hyaline, granular, fatty, waxy and epithelial casts, renal epithelial cells undergoing extensive fatty degeneration, free fat globules, leucocytes and a few red blood cells. If an acute exacerbation of chronic diffuse nephritis occurs, then there is, in addition to the above, more blood cells and blood casts in the urine, the quantity is more diminished and the œdema is much more extensive.

Chronic Interstitial Nephritis (Granular Contracted Kidney)—

The urine is increased in quantity, especially at night. It is pale in color. Specific gravity is diminished, usually between 1,010 and 1,004.

Albumen is present in traces, but at times may not be found. The morning urine may be free from albumen, but the evening urine may contain it. It is necessary to centrifugalize the urine to obtain sediment, and even then it is small in amount.

Microscopical examination shows a small number of hyaline and finely granular casts, an occasional epithelial cell and a few leucocytes.

Renal Tuberculosis—

Early in the case polyuria is of frequent occurrence. Often the passage of blood is the first thing to attract the attention of the patient.

The blood may be microscopic only, or macroscopic. It is usually intermittent, the average length of the bleeding is three days. The blood and the urine are intimately mixed. Pus is frequent in all cases in which the pelvis of the kidney is involved. Amount is very variable, from a few leucocytes to ¼ the volume of the urine. As a rule this pyuria is constant.

Albumen is of course always demonstrable whenever blood or pus is present—the latter by itself is only responsible for a trace; casts may be present, but they are not a constant factor in the urinary picture, while the association of hematuria and pyuria with an acid urine should always excite suspicion.

The diagnosis of renal tuberculosis demands the demonstration of the tubercle bacillus as well.

Arteriosclerotic Kidney—

Urinary findings the same as in chronic interstitial nephritis.

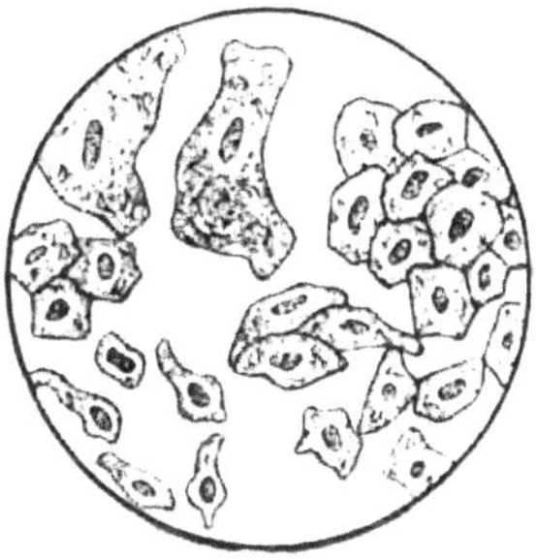

Fig. 5.—Epithelial cells from bladder and urethra.

Fig. 6.—Epithelial cells from kidney.

Fig. 7.—Pus cells.

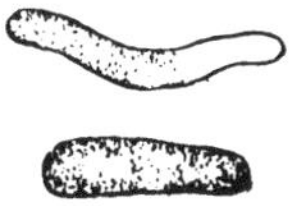

Fig. 8.—Fine granular casts.

Fig. 9.—Coarse granular casts.

Fig. 10.—Hyaline casts.

Fig. 11.—Waxy casts.

Fig. 12.—Leucocyte or pus casts.

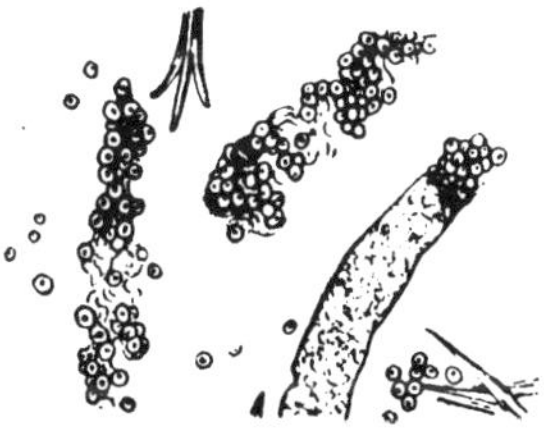

Fig. 13.—Free red blood cells and blood casts.

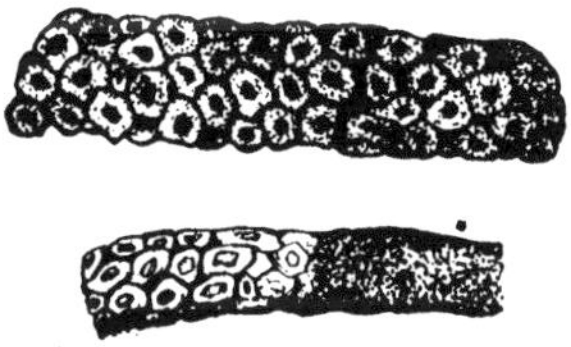

Fig. 14.—Epithelial casts.

Fig. 15.—Fat casts.

Fig. 16.—Mucous threads.

CHAPTER V

GASTRIC CONTENTS

In an examination of stomach contents the material from a fasting stomach and also the contents following a test meal should both be examined.

It is best to obtain the contents of a fasting stomach early in the morning, as in normal stomachs no food should be present; if it is present it is a sign of stasis. If as much as 50 c.c. of fluid is obtained it indicates hypersecretion or stasis. After removing the fasting contents the test meal should be given.

Test Meal

Ewald Test Meal—

This consists of one large slice of bread and a glass and one-half of water. Masticate the bread thoroughly. This should be removed in one hour. Pour in 200 c.c. of water and remove this as the lavage.

An objection to this test meal (Ewald's) is the presence of a certain amount of lactic acid in the bread. A good substitute for the bread is one Shredded Wheat Biscuit. This is known as Dock's test meal.

Boas' Test Meal—

Boil one tablespoonful of oatmeal in 800 c.c. of water till the volume is reduced to about 400 c.c. Administer to patient and remove in one hour.

Physical, Chemical and Microscopical Examination

A physical, chemical and microscopical examination should be made of the fasting contents and the test meal removed in one hour.

Physical Examination

Quantity—

Measure amount of fasting contents, and the amount after a test meal, which should be about 100 c.c., if 200 or 300 c.c., either motor insufficiency or hypersecretion is probable. Note proportion of fluid in the total quantity of contents; the food residue should be about one-fourth.

Color—

Coffee-ground appearance suggests cancer. Reddish color is due to blood and suggests ulcer. Green or yellow color is due to bile.

Odor—

The odor of butyric acetic and lactic acid and yeast is characteristic of fermentative changes in the stomach.

Mucous—

Large amounts suggest chronic gastritis. Small amounts occur normally.

Chemical Examination

Qualitative examination should be made for the following: Free acids, combined hydrochloric acid, free hydrochloric acid, organic acids (lactic acid), pepsin, rennin and blood.

1. Free Acids—

Use Congo red paper. A blue color indicates the presence of free acids.

2. Free Hydrochloric Acid—

Topfer's Test.—Add to 5 c.c. of gastric contents two drops of Topfer's solution. A carmine red color indicates free hydrochloric acid.

Gunzburg's Test.—Mix a drop of this reagent with a drop of gastric contents in a porcelain dish or on a glass slide. Evaporate slowly over a flame. A bright red color at contact indicates free hydrochloric acid.

3. Combined Hydrochloric Acid—

When there is no free hydrochloric acid present it is important to know whether there is achylia gastrica (no free nor combined hydrochloric acid) or merely hypo-acidity (combined hydrochloric acid being present).

Take 10 c.c. of gastric contents. Add a pinch of barium carbonate, which is insoluble. Evaporate in a porcelain dish to dryness and fuse at a low red heat.

If hydrochloric acid combined with proteids is present, the heat will liberate the hydrochloric acid and the chlorine of the hydrochloric acid will unite with the barium of the barium carbonate and form barium chloride. Allow this to cool and add distilled water to the residue of the porcelain dish. When it is dissolved, filter. Divide the filtrate into two parts. Add to one part a few drops of a saturated solution of sodium carbonate.

A precipitate, which is barium carbonate, indicates combined hydrochloric acid in the original. The other part of the filtrate should be used as a control.

4. Organic Acids—

Lactic.—Dilute an aqueous solution of ferric chloride with water to a faint yellow color. Fill two test tubes about one-third full with this solution. Use one tube as a control, and to the other add 5 drops of gastric contents; an intensification of the yellow color indicates lactic acid.

Uffelmann's Test for Lactic Acid.—Into a test tube put about ½ inch depth of 5 per cent carbolic acid. To this add one drop of an aqueous solution of perchloride of iron; a deep amethyst color results. Dilute this with distilled water until it can be seen through readily. Into a test tube ½ filled with this diluted reagent pour 5 to 8 drops of the gastric juice. The amethyst color changes to a canary yellow in the presence of lactic acid.

5. Pepsin—

Take 10 c.c. of unfiltered stomach contents in each of two tubes and to each tube add some albumen of a hard-boiled egg. Add to one of the test tubes an equal amount of .4

solution of hydrochloric acid, and to the other an equal amount of water. No free hydrochloric acid should be present at the beginning of the experiment. Place each in thermostat for 24 hours and note the digestion of the egg albumen, which means that pepsin is present.

6. Rennin—

Take 10 c.c. of unfiltered stomach contents and neutralize to litmus by adding 10 per cent sodium hydrate solution. Take 10 c.c. of milk and boil to kill bacteria. Allow this to cool. Add 5 c.c. neutral gastric contents to the 10 c.c. of milk and put in water bath at body temperature for 30 minutes. A curdling of the milk indicates the presence of rennin. It is well to take varying amounts of the gastric contents, as this will indicate the quantity of the ferment.

7. Blood—

A. *Guaiac Test.*—Place 10 c.c. of gastric contents in a test tube and add 2 c.c. of glacial acetic acid and 15 c.c. of ether. Insert a cork and shake gently for several minutes. After the ether has separated, decant. Add to the etheral solution 10 drops of freshly prepared tincture of guaiac and 2 c.c. of hydrogen peroxide. A blue color indicates the presence of blood.

B. The following chemical test may also be used. Place some of the material supposed to contain blood on a glass slide, add to this a minute crystal of sodium iodide, and then two drops of glacial acetic acid. Cover with cover glass and warm gently over flame until bubbles appear. If blood is present, crystals of hemin will easily be seen under high dry lens.

Quantitative Examination

Quantitative examination should be made for *Free Hydrochloric Acid, Total Acidity, and Combined Hydrochloric Acid.*

Free Hydrochloric Acid—

To 10 c.c. of unfiltered gastric contents add 3 drops of Topfer's solution. Titrate with decinormal sodium hydrate solution until the disappearance of the carmine red color. This point represents the neutralization of the free hydrochloric acid in the contents used.

To estimate the quantity of free hydrochloric acid, multiply the number of c.c. of the decinormal sodium hydrate solution used in the titration by 10. This gives the amount of free hydrochloric acid in 100 c.c. of gastric contents in terms of decinormal sodium hydrate. The result may be expressed in per cent of hydrochloric acid if the above quantity is multiplied by .00365, i. e., the quantity of hydrochloric acid which is neutralized by 1 c.c. of decinormal sodium hydrate. The normal quantitative values of free hydrochloric acid vary between .07 and .18 per cent, or 20 to 50 c.c. of decinormal sodium hydrate per 100 c.c. of gastric contents.

Total Acidity—

To the same contents, in which the free hydrochloric acid has already been neutralized, add 3 drops of a 1 per cent alcohol solution of phenolphthalein. Continue the titration with decinormal sodium hydrate solution until a permanent red color is obtained. This represents the neutralization of all the acid constituents of the gastric contents (free mineral, organic acids and combined acids).

To estimate the total acidity, multiply the number of c.c. of the decinormal sodium hydrate solution used from the beginning of the titration up to this point by 10. This gives the total acidity of 100 c.c. of gastric contents in terms of decinormal sodium hydrate. The result may be expressed in per cent hydrochloric acid by multiplying the above quantity by .00365. The normal quantitative values of the total acidity vary between .15 and .30 per cent, or 40 to 80 c.c. decinormal sodium hydrate per 100 c.c. of gastric contents.

Combined Hydrochloric Acid—

To 10 c.c. of gastric contents add 3 drops of a 1 per cent solution of phenolphthalein. Titrate with decinormal sodium hydrate solution until a pink color is obtained throughout. Read off from the burette the number of c.c. of decinormal solution required. Take another 10 c.c. of the gastric contents and add 3 or 4 drops of a 1 per cent solution (aqueous) of alizarin as an indicator. Titrate with decinormal sodium hydrate solution until the yellow color has disappeared completely and a purple color appears. Read off from the burette the number of c.c. of the decinormal sodium hydrate solution required. Subtract the number of c.c. required with alizarin as an indicator from the number required with phenolphthalein as an indicator, and multiply by 10, and this will give the amount of combined hydrochloric acid expressed in terms of decinormal sodium hydrate solution. The indicator, phenolphthalein, reacts to free acids, acid salts, and combined acids, but the indicator, alizarin, reacts only to free acids and acid salts, but not combined acids.

Microscopical Examination

This should be done in all fasting contents, which should not be filtered. Examine for: (a) food particles, as muscle fibers, starch cells, starch granules and fat globules, (b) mucus, (c) red blood cells, (d) leucocytes, (e) sarcinæ, (f) yeast cells, (g) bacteria, as Oppler-Boas bacillus.

Stomach Contents in Disease

Gastric Cancer—

Amount varies depending on whether or not there is a stenosis.

Blood is present when there is ulceration, which may be early or late in the disease. The color of the stomach contents is not changed by small quantities of blood, and an adequate chemical examination is necessary for the detection. If much blood is present, and remains in the stomach for a time, a coffee-ground appearance is given to the stomach contents.

In about 80 per cent of the cases there is no free hydrochloric acid present; however, when a malignant growth begins on the base of an old ulcer the hydrochloric acid may not only be normal in quantity, but even excessive in amount.

When free hydrochloric acid is absent there is usually always a certain amount of lactic acid present.

Microscopically one often finds the Oppler-Boas bacillus when lactic acid is present, and red blood cells may also be found.

Gastric Ulcer—

Seventy-five per cent of all cases show the presence of blood, which may be in macroscopical amounts, or a chemical examination may be required to demonstrate its presence. Free hydrochloric acid is increased in about half the cases, as is also the total acidity. Blood may be present in feces.

Gastric Neurosis—

The findings are variable. Hydrochloric acid is sometimes increased and sometimes diminished. Pepsin is normal.

Chronic Gastritis—

In early cases hydrochloric acid may be increased. It is generally diminished in well marked cases, and is often absent in advanced cases. When absent, lactic acid is present in traces, mucus is present and is significant of the disease; motility and absorption is generally deficient.

Achylia Gastrica—

This is frequently the terminal stage of chronic gastritis. There is an entire absence of hydrochloric acid, and a very low total acidity. A small amount of lactic acid may be present. The motility of the stomach is fairly good. This condition is often associated with gastric carcinoma and pernicious anemia.

CHAPTER VI

BLOOD

Examination of a Drop of Fresh Blood—

Secure a drop of blood on a cover slip and drop same on a slide and immediately examine microscopically. The slide and cover slip must be absolutely clean to enable the blood to spread in a very thin layer. Note size, shape and color of red corpuscles. Note the leucocytes and the ameboid movements of the polynuclear variety. Note the fibrin. One may examine fresh blood for the malarial parasite, embryos of filaria sanguinis hominis and the trypanosoma gambiense.

Hæmoglobin—

For an accurate estimation use the Sahli apparatus. The hæmoglobin is expressed in percentage, 100 per cent being considered as the normal.

Estimation of Hæmoglobin—

By the Use of the Sahli Hæmometer.—Fill the graduated tube to mark 10 with N/10 hydrochloric acid. Fill the pipette up to the mark 20 cu. m.m. with blood. This is quickly discharged into the N/10 hydrochloric solution. Shake and let stand for one minute. Now dilute with water until color matches that of the standard solution. The height of the column is then read. This gives the hæmoglobin percentage.

By the Talquist Scale of Colors.—A rather large drop of blood is collected on one of the squares of filter paper that is supplied in the book. The gloss is allowed to disappear, and it is then placed under the perforation in one of the red strips. It is moved until the color of the drop of blood corresponds with one of the shades of red. This represents the obin percentage of the blood. The results are inexact, but suffice for rapid bedside work.

Color Index—

This is the quotient obtained by dividing the percentage of hæmoglobin by the percentage of red blood corpuscles, 5,000,000 red cells per cu. m.m. being considered as 100 per cent

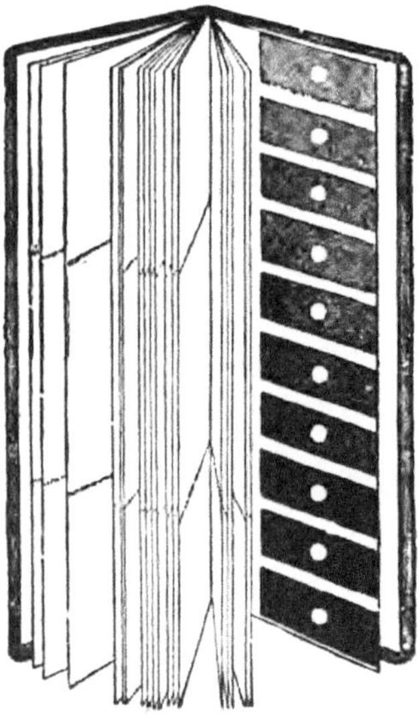

Fig. 17.—Talquist hæmometer.

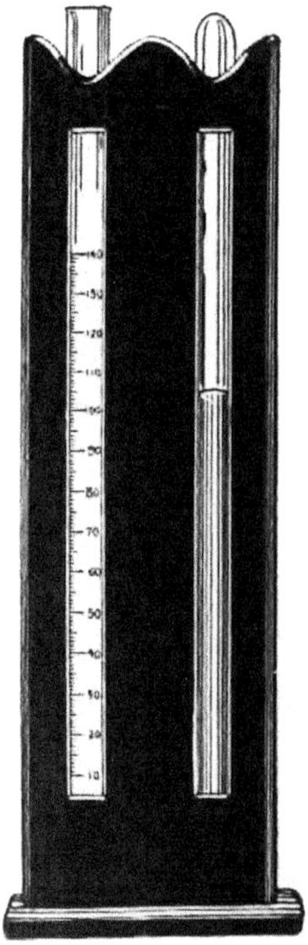

Fig. 18.—Sahli hæmometer.

of the corpuscles. Normally the color index is about one. When the index is less than one, it indicates that the average corpuscle is poor in coloring matter, whereas with a high color index the corpuscles are rich in hæmoglobin.

Making a Blood Smear

A blood smear for purposes of staining is made on either cover slip or slide. For delicate and accurate work the former is superior, but for all practical purposes the latter suffices.

The slide or cover slip intended for the smear should be first thoroughly cleansed with soap and water, rinsed in clear water, and finally in about 50 per cent alcohol. It should be rubbed dry with a clean cloth, care being used not to touch the flat surface with the fingers, but only the sides.

Cleanse the lobe of the ear or end of the finger with soap and water, and then alcohol. Usually alcohol alone will suffice. Wipe dry and pierce with a surgical needle. Wipe away the first few drops of blood. Touch the center of the cover glass against the top of the blood drop and immediately drop this on top of a clean cover slip. About the time the blood ceases to spread draw them apart, but keep their surfaces parallel.

A smear may be made on a clean slide by getting a drop near one end and with another slide gradually drawing this drop over the slide.

Blood Counting

For the purpose of counting erythrocytes or leucocytes, a counting chamber and pipette are required. The pipette for red cells and white cells are graduated differently in order to make the dilution satisfactory for the count, the red cells requiring much greater dilution than the white cells. The pipette for the red cells is marked 0.5 at half the distance of the capillary tube and 1 at the upper end of the tube. Above the bulb is the mark 101. If the blood is drawn only to .5, as is customary, and then the diluting fluid to the mark 101, then the red cells have a dilution of 1 to 200. If the blood is drawn to 1 and the diluting fluid to 101, then the dilution is 1 to 100.

The pipette for the leucocytes is marked 0.5 at half the distance of the capillary tube, and 1 at the upper part of the capillary tube. Above the bulb is the mark 11. If the blood is drawn only to .5, as is customary, and then the diluting fluid to the mark 11, then the leucocytes have a dilution of

1 to 20. If it is drawn to 1 and the diluting fluid to 11, then the dilution is 1 to 10.

The counting chamber consists of a thick glass slide, on the center of which is mounted a small circular glass disc

Fig. 19.—Red cell pipette.

Fig. 20.—White cell pipette.

which is ruled according to either Thoma, Türck, or Zappert-Ewing. This circular ruled disc is surrounded by a square glass table mounted also on the slide. The glass table is exactly 0.1 m.m. above that of the ruled disc. A moat separates the table from the ruled disc.

The disc ruled according to Thoma is as follows: This consists of five square millimeters. The central square millimeter, which is used for counting the erythrocytes, is subdivided into 400 small squares. By means of double lines these smallest squares are grouped into blocks of 25, a convenient unit to employ in counting the red cells. For the leucocytes all five square millimeters are counted and the average in one square millimeter estimated.

The rulings according to Türck and Zappert-Ewing are similar to that of Thoma, but in the latter two there are nine

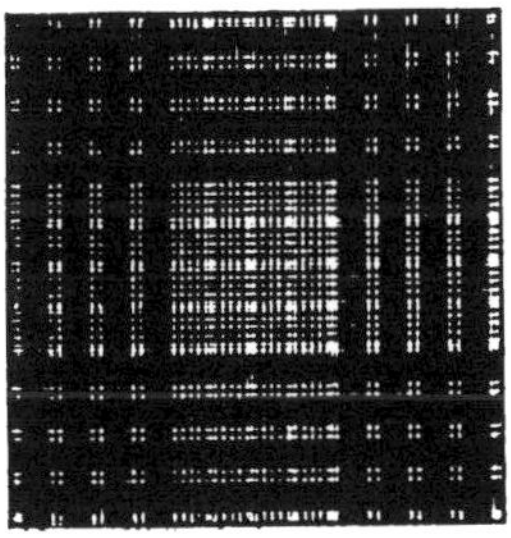

Türck.

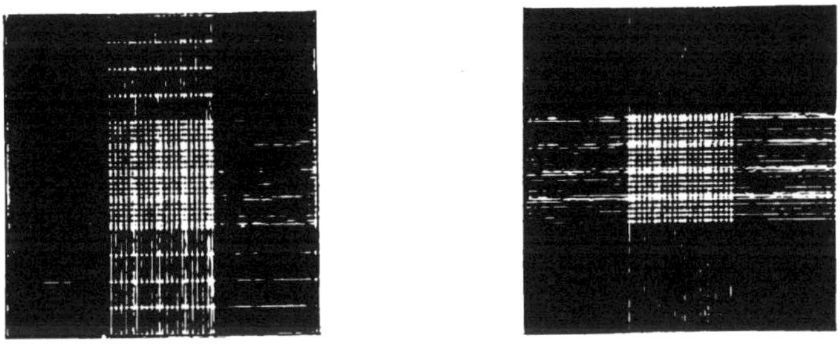

Zappert-Ewing. Thoma.

Fig. 21.—Blood counting chambers.

square millimeters instead of five. These latter two differ in ruling of the four corner squares. For the red count only the central square millimeter with its 400 small squares is used, while for the leucocytes all nine are counted and the average taken.

The counting chamber ruled according to Türck is the most serviceable and is the one the author recommends.

1. Red Corpuscles—

Draw the blood into the blood pipette up to the mark .5 and dilute with Hayem's solution up to the mark 101. Mix thoroughly. Blow out two or three drops, then place a drop in the center of the ruled glass piece of the counting slide, and gently place on the cover slip. If a small amount passes out into the moat it will not matter, but it should not pass out under the cover slip, as this raises it to a greater distance than .1 of a millimeter. Examine for Newton's rings. Count the red corpuscles in 25 small squares at each of the four corners of the central ruled square, then multiply by four, then by ten, and finally by the dilution 200. Repeat the count and take the average. In making the count, cells are often seen in the position across the lines, so the rule is to count the cells crossing the left and lower lines and not counting those touching the upper and right lines. This count gives the number of red blood cells in a cubic millimeter of blood.

2. White Corpuscles—

Draw the blood into the blood pipette up to the mark .5 and dilute with 1 per cent acetic acid up to the mark 11. Mix thoroughly. Blow out two or three drops, then place one drop in the center of the ruled glass piece of the counting slide, and gently place on cover slip. If a small amount passes out into the moat it will not matter, but it should not pass out under the cover slip, as this raises it to a greater distance than .1 of a millimeter. Examine for Newton's rings. Count the leucocytes in the several square millimeters, and get the average in one. Multiply this by 10, then by 20. Repeat this and get the average.

Cleaning the Blood-counting Apparatus—

(a) *Counting Chamber.*—Clean this with water only. A little soap may be used. Never use alcohol, ether or other solvent, as the cement by which the ruled disc is fastened to the slide may be dissolved.

(b) *Pipettes.*—Always clean immediately after using. The following steps should be employed: (1) Blow out contents. (2) Draw up into it distilled water and blow it out. (3) Repeat this. (4) Draw up 95 per cent alcohol once or twice and blow it out. (5) Repeat with ether, the last time removing the rubber tubing and blowing out through the large end.

Method of Staining Blood Smears—

1. Let the specimen dry in the air.
2. Cover well with Wright's stain for one minute.
3. Add distilled water, drop by drop, until a delicate metallic scum appears on the surface, usually 3 drops.
4. Leave this dilute stain on for 2 minutes.
5. Wash in distilled water about 5 seconds, or until the preparation has a pinkish color.
6. Dry quickly, mount and examine.

Examine the red blood cells and note variations in size (mikrocytes and makrocytes), and shape (poikilocytes). Note loss of color (achromia). Stippling; i. e., appearance of granules in the cell. Polychromatophilia. Note the appearance of blasts, either normoblasts or megaloblasts.

Examine the white blood cells (leucocytes). Make a differential count by counting at least 200 white corpuscles.

Note the number and percentage of:

1. Polynuclear basophiles (mast cells).
2. Polynuclear neutrophiles.
3. Polynuclear eosinophiles.
4. Small lymphocytes.
5. Large lymphocytes.
6. Large mononuclears.
7. Transitional.

Examine for myelocytes (neutrophilic, eosinophilic, or basophilic). Examine for myeloblasts.

Examine the blood plates. These appear with Wright's stain as round or oval bodies stained purplish. They are often found in clumps and are about one-fourth the size of a red blood cell.

Examine for parasites.

Malarial Parasites.—Three types are found: tertian, quartan, and æstivoautumnal. The early non-pigmented forms of all three are so similar that a positive differential diagnosis is very difficult. The later stages, however, are quite characteristic. Stained with Wright's stain they appear as follows:

Tertian.—Infected cell appears larger than the normal red cell and is paler. The parasite has a blue body and brown pigment granules scattered more or less throughout the body of the parasite.

Quartan.—Infected cell appears smaller than the normal red cell. The parasite has a blue body in which is found coarse brown pigment granules. The granules tend to collect near the periphery of the organism.

Æstivoautumnal.—Diagnosis made by the appearance of crescents and ovoids in the peripheral blood. Crescents which appear blue are slightly larger than the red cells and have a distinct crescentic shape. They often show a fringe of the degenerated red cell on the concavity of the crescent. Pigment is found near the center of the crescent. The crescents may become ovoid in shape, called ovoids, or spherical in shape, called spheroids.

Staining of Thick Blood Smear—

A large drop of blood (20 cu. m.m.) is placed on cover slip or slide and partially spread out by the usual method. Fix by covering smear with a 2 per cent formalin solution containing ½ to 1 per cent of glacial acetic acid. Let this remain 1½ minutes. This extracts hæmoglobin from the red blood cells at the same time as the specimen is fixed. Cover specimen with dilute sodium carbonate solution, 2 to 3 per cent, for 1 to 2 minutes, wash off gently with water. Blot dry. Then stain in the usual way with Wright's blood stain. This is useful in finding the parasites when their number in the blood is small.

Widal's Serum Reaction—

Use a bouillon culture of bacillus typhosus 12 to 24 hours old. Examination by high dry lens should show the bacilli in active motion and unclumped. A growth on agar agar about 24 hours old may also be used.

Get the blood from the patient in a glass tube drawn out at both ends into a capillary tube, or from vein at bend of elbow with glass syringe. Centrifugalize to throw corpuscles to bottom. Drop 9 drops of physiological salt solution into a small-size glass tube, and 19 drops into another. Add a drop of the serum to each tube and mix thoroughly. Place a drop of this diluted serum from each tube on two cover slips and add a platinum loop of typhoid bacilli grown on bouillon to each drop. This gives a dilution of 1 to 20, and 1 to 40 respectively. Place cover slip on a hanging glass slide and examine microscopically with a high power dry lens.

The serum reaction is regarded as positive when there is complete clumping of the bacilli and absolute cessation of motility. The time limit for the test is one hour, although the agglutination often occurs within a few minutes. A reaction at 1 to 20 is very suggestive, while a reaction at 1 to 40 can be accepted as conclusive evidence of typhoid infection.

Another method of securing the blood is by getting a large drop on a glass slide and extracting the blood with water. Place 5 or 6 drops of water on a large drop of blood and let stand for 15 minutes. Then gently stir with a platinum loop, but do not mix up the cells and fibrin in the extract. Take one platinum loop full of this extract and add it to the same quantity of a 12 to 18 hour bouillon culture of typhoid bacilli. This gives an approximate dilution of 1 to 40. Examine this for agglutination.

Blood Culture—

This should be done in all cases of suspected septicemia. The diagnosis of typhoid fever is earliest made by the detection of typhoid bacilli in the blood. The best culture media for the typhoid and colon bacilli is sterilized ox bile or, perhaps better is ox bile containing 10 per cent glycerine and 2 per cent peptone. Flask containing 25 to 100 c.c. of the culture media is customarily employed.

In suspected typhoid fever the following is carried out: The flexor surface of the elbow is washed with soap and water or alcohol, then a large area is painted with tincture of iodine. The vein is made to stand out well by bandaging arm above elbow. Use sterile glass syringe (Luer model) and insert needle parallel with the vein and in the direction of the blood current. Draw into syringe 6 or 8 c.c. of blood. Have at least three flasks of the bile media. In flask No. 1 place 1 c.c. of blood, in No. 2 place 2 c.c. of blood, and in No. 3 place 3 or 4 c.c. of the blood. Incubate for 36 hours, then inoculate a bouillon and also an agar agar tube, and in 36 hours note presence or absence of a growth.

In a case of suspected septicemia a very serviceable culture media is glucose bouillon in flasks, each flask containing 50 to 150 c.c. of the culture media. Glucose-agar in rectangular-shaped bottles is also very good. Two flasks of glucose bouillon, and also two bottles of glucose-agar should be used. The agar should be melted and allowed to cool to 42° C. before inoculating. In flask No. 1 (glucose buillon) place 1 c.c. of blood, and in flask No. 2 place 3 to 5 c.c. of blood. In bottle No. 1 (glucose agar agar) place 1 c.c. of blood, and in bottle No. 2 place 3 to 5 c.c. of blood. Shake, then lay on side and allow to harden. Incubate both at 37° C. for 24 to 48 hours. The agar culture should be placed with culture media lying above. From the bouillon, after 24 to 36 hours, reinoculate other culture tubes. Note after 24 hours presence or absence of growth. Blood is obtained in same manner as in suspected typhoid fever. Extreme care should be used in inoculating so as to avoid contamination.

Changes in the Blood in Various Diseases

Pernicious Anemia—

Number of red blood cells usually greatly diminished. The percentage of hæmoglobin also diminished, but not to the same extent as the number of red blood cells. Color index is therefore high, about 1. or 1+. Very many macrocytes (usually well stained), and many poikilocytes, are present. Often there is seen granular degeneration, also polychromatophilia. Variable number of nucleated red blood cells, the predominating variety being the megaloblasts, although normoblasts are often seen. Leucocytes in the majority of cases are somewhat diminished with relative increase of the lymphocytes.

Acute Lymphatic Leukemia—

Number of red blood cells more or less diminished. The percentage of hæmoglobin also diminished and more in extent than the red blood cells. This gives a low color index. Very often nucleated red blood cells, which are chiefly normoblasts, are seen. Leucocytes are more or less markedly increased, from 50,000 to 250,000 per cu. m.m. The predominating cell is the lymphocyte, the number of which frequently exceeds 80 per cent, the majority being the large lymphocyte.

Chronic Lymphatic Leukemia—

Number of red blood cells diminished. The percentage of hæmoglobin is diminished to a greater extent than the number of red blood cells. The color index is therefore low. Nucleated red cells are not so often seen as in the acute variety, but they are frequently found. Leucocytes are very much increased, the average being about 350,000 per cu. m.m. The variety especially involved is the small lymphocytes, more than 90 per cent being of this variety.

Spleno-Myelogenous Leukemia—

Number of red blood cells more or less diminished. The percentage of hæmoglobin is also diminished, and to a greater extent than the red blood cells. The color index is therefore low. Many poikilocytes, stipple cells, macrocytes and microcytes are present. Very many nucleated red cells are present. These are chiefly normoblasts, but megaloblasts are also found. Leucocytes are very much increased in number, on an average of 350,000 to a cu. m.m. of blood, but the number runs from 150,000 per cu. m.m. to 600,000 or 700,000. The type chiefly involved is the myelocyte, 40 to 60 per cent of the leucocytes being of this variety. The neutrophilic, eosinophilic and basophilic myelocytes are all present, but the largest per cent is of the neutrophilic variety.

Chlorosis—

The hæmaglobin is markedly diminished; the number of red blood cells very slightly so and sometimes not at all. Color index is always low. Achromia is marked, poikilocytes, stipple cells, and occasionally a few normoblasts. The leucocytes are not affected as to the number. There is usually a slight relative increase in the lymphocytes. This condition occurs at about the age of puberty in young girls who are very nervous.

Splenomegaly or Splenic Anemia—

A condition characterized by enlargement of the spleen, an anemia of a secondary type, without leucocytosis or lymphatic enlargement, and a gradual downward course. If there is, in addition to the enlarged spleen, cirrhosis of the liver, jaundice and ascites, the condition is called Banti's disease.

CHAPTER VII

SEROUS FLUIDS

This includes both transudates and exudates. Transudates are non-inflammatory. The specific gravity is 1008 to 1018, with only a few cells and a small amount of albumen; i. e., .2 to 2 per cent. Exudates are inflammatory in origin. The specific gravity is 1018 to 1026, larger number of cells and a greater abundance of albumen, 2 to 6 per cent.

Serous Fuids

The Serous Fluids are: 1. Pleural. 2. Peritoneal. 3. Pericardial. 4. Cerebrospinal. In examination of serous fluids the methods that apply to one will apply to all.

Physical Examination—

Note the color, turbidity, and take the specific gravity. Note the amount of fibrin or clot, take reaction.

Chemical Examination—

Qualitative and Quantitative tests for albumen. (See Urine for tests.)

Cytodiagnosis—

This is important as a diagnostic method and consists in a differential count of the cells in a transudate or exudate. The following should be carried out:

1. Place fluid in centrifuge tubes and centrifugalize for 3 minutes.
2. Pour off supernatant fluid.
3. Make smear from sediment in same manner as blood smear.
4. Let this dry in the air.

5. Cover the slip with Wright's stain for 45 seconds and then add 3 drops of water and let remain for 2 minutes.
6. Wash rapidly in distilled water.
7. Quickly blot dry, using filter paper.
8. Mount and examine with an oil-immersion lens, and count the different types of cells.

A predominance of polymorphonuclear leucocytes points to an acute infection.

A predominance of lymphocytes usually means tuberculosis, but may mean parasyphilis or cerebrospinal syphilis.

A scarcity of cells, with a predominance of endothelial cells, indicates a transudation.

Carcinoma of the serous membrane shows a predominance of endothelial cells, but with large numbers of lymphocytes and red blood cells.

Bacteria in Serous Fluids—

Pour a few c.c. of the fluid into a flask of nutrient bouillon. Incubate for 24 to 36 hours and then make a smear on cover slip and stain with Löffler's methylene-blue or Gram's stain. The technique of these staining processes has already been given.

Make smear also as in the method for cytodiagnosis and stain with Löffler's blue and Gram's stain. In this way bacteria are found.

Total Cell Count of Serous Fluids—

Draw up to the mark I in the white cell pipette a 4 per cent solution of glacial acetic acid tinged slightly with a few drops of a solution of gentian violet. Now draw up the serous fluid to the mark 11. Shake well, and blow out first 2 or 3 drops. Place a drop on the center of the blood counting chamber, gently place on cover slip, examine for Newton's rings, count the cells in the different square millimeters on the counting chamber. Take the average in one square m.m. Take 1/10 of this and add back to the number. Now multiply by 10. This gives total number of cells in cubic m.m. of the fluid. In spinal fluid the number is normally from 2 to 10.

In acute infections of the meninges the number may run up into the thousands. In tuberculous meningitis the number runs from 150 to 250 per cu. m.m.

In parasyphilitic lesions and cerebrospinal lues the number runs from 24 to 150 per cu. m.m. In the latter two diseases there is a very high percentage of lymphocytes, while in acute infection of the meninges there is a high percentage of polymorphonuclear neutrophiles.

Noguchi's Butyric Acid Test for increase in the globulin content of the cerebrospinal fluid. This test is based upon the observation that in syphilitic and parasyphilitic affections of the central nervous system the globulin content is increased.

The test is as follows:

Take .3 c.c. of the meningeal fluid, absolutely free from blood, and add to it 1 c.c. of 10 per cent solution of butyric acid in physiological saline. Boil this a few seconds over a flame, then quickly add .2 c.c. normal sodium hydrate solution and boil for a few seconds longer. In the presence of an increased globulin content a granular or flocculent precipitate appears. This gradually settles to the bottom of the tube. Usually 2 hours is given for the appearance of this granular precipitate.

Nonne-Apelt Test for Determining Protein in Cerebrospinal Fluid.—Two c.c. of cerebrospinal fluid are mixed with an equal quantity of a neutral saturated solution of ammonium sulphate (Merck). Compare this after 3 minutes with another tube containing only cerebrospinal fluid. If there is no difference, or only a very faint opalescence, the reaction is considered negative. If there is a distinct opalescence or a turbidity, the reaction is said to be positive. This is considered as significant of globulin excess.

Lange Colloidal Gold Method—

Four solutions necessary: (1) 1 per cent solution of chloride of gold in distilled water—use absolutely clean utensils. (2) 2 per cent solution of potassium carbonate. (3) 1 per cent solution of formaldehyd. (4) 0.4 per cent solution of sodium chloride, fresh.

Preparation of the Gold Solution.—Place into a perfectly clean 1000 c.c. flask 500 c.c. of fresh doubly distilled water. Add to the 500 c.c. of distilled water 5 c.c. of the carbonate solution, and in about 30 seconds 5 c.c. of the gold chloride solution. Bring quickly to a boil. As soon as bubbles appear remove from over flame and add 3 ¾ c.c. of the formalin solution. Shake in a rotary manner until fluid is a cherry red color. The fluid should be absolutely clear.

Test—Have 10 perfectly clean and clear test tubes in a rack. Place in test tube No. 1 exactly 1.8 c.c. of the 0.4 per cent salt solution; in the remaining tubes 1 c.c. of the 0.4 per cent salt solution. Into test tube No. 1 place 0.2 c.c. of spinal fluid. Mix well. Remove 1 c.c. and place in tube No. 2. Mix this and place 1 c.c. into tube No. 3. Continue this manner of dilution to tube No. 10. Reject 1 c.c. from last tube. This gradually dilutes the spinal fluid. Now add to each tube 5 c.c. of the gold colloidal solution and mix well at once. Let stand for 24 hours at room temperature, then examine.

With normal cerebrospinal fluid no change of color occurs in any of the tubes. If only a single tube shows color change, it indicates faulty technique.

With pathological cerebrospinal fluids there is a change of the red color, in two or more tubes, to a more or less marked blue color. Any form of syphilis of the central nervous system is indicated by varying degrees of color change. These changes are noticeable in tubes 3, 4, and 5. As a rule, fluid from general paresis and cerebrospinal syphilis give stronger reactions than that of tabes. In this test the cerebrospinal fluid should be entirely free from blood.

Technique for the Performance of Lumbar Puncture

Locate the intervertebral space on a line with the crest of the ilium. This is usually the space between the third and fourth lumbar vertebra. Cleanse with soap and water an area about the size of the hand. Rub this area with alcohol, and then it may or may not be painted with tincture of iodine. Have patient on left side near edge of bed, with head and shoulders slightly elevated on a pillow, and legs flexed well at the hips. With a sterile lumbar puncture needle, 6 to 8 centimeters in length, enter as near the center of the space between the spinous processes of the third and fourth lumbar vertebræ as possible. (The author prefers to enter in the mid line, but many others prefer to enter about 1/5 to 2/5 of an inch to the right or left of the mid line.) Pass the needle about the distance of 4 centimeters with point inclined slightly toward the head, and if it strikes bone withdraw a short distance and change slightly the direction of the point of the needle, and continue this until you feel it pass the dense dura mater into the spinal subarachnoidean space. Withdraw the stilette and the spinal fluid will immediately begin to drop out in rapid intermittent drops. If pressure is very much increased it may escape in a stream. The fluid should be collected in sterile test tubes. One should use at least three tubes, as the first will likely contain a little blood, perhaps also the second tube, while the third tube will very probably contain no blood. The presence of blood interferes with certain of the tests.

CHAPTER VIII

INTESTINAL CONTENTS

A *Macroscopic, Chemical* and *Microscopic* Examination of the stool should be made whenever intestinal trouble is suspected. Especially is this true in the South.

The examination of the stools is very much simplified if the patient's diet is restricted for two or three days before. The diet should consist of milk, toast, gruel, eggs, oatmeal, butter, rare steak and potatoes. If this is not done, then ask the patient what he has been taking for the past two or three days.

Macroscopic Examination

Note the color, quantity, frequency, consistency, odor, and presence of mucus, blood, pus, curds, round worms, hook worms, segments of tape worms, etc.

Color—

The normal brown color is due to hydrobilirubin. Infants' stools are normally bright or golden yellow. The color of the stools varies with: (a) Food—light with milk and bread, dark with blackberries, red with wine and exclusive meat diet, etc., green with green vegetables. (b) Drugs—green after administration of calomel, black after bismuth subnitrate and iron. (c) Blood—if from stomach or duodenum, tarry stools; if from colon or rectum, bright red. (d) Bile—clay-colored from diminished secretions, obstruction to flow of bile, and from unabsorbed fat. (e) The green color of stools is pathological, and is due to the presence of bilirubin.

Quantity—

This depends on amount and character of diet, and habits of patient.

Frequency—

Stools may be numerous but without fæcal material. Frequent stools that come from colon are usually small; if from small intestines, usually large.

Consistency—

This depends on how long the fæces has remained in the rectum. Frothy stools indicate intestinal fermentation.

Odor—

This is important in infants. Normally it is slightly sour, foul in proteid putrefaction, sour in acid fermentation, odorless in cholera infantum.

Mucus—

This is usually indicative of inflammation of the large intestines. If it comes as a coating to the fæces, or as membranous flakes, shreds or gelatinous clumps, it is from the large intestines. If it is intimately mixed with the stools, it is from the small intestines.

Blood—

See under heading of "Color."

Pus—

If in large quantities, is due to an abscess rupturing into intestines, or to ulcerative colitis. If leucocytes come from above cœcum they will be digested, and only the nucleus will be seen.

Curds—

These are often seen in infants' stools and are an indication of imperfect proteid digestion.

The stools should always be examined also for worms or their segments. This is best done by taking a portion of the fæces and adding a little water until the stool is a thick liquid, at the same time mixing the two together with a glass rod or pestle, then place on a dark glass. Note also the presence of mucus or undigested food.

Chemical Examination

Take reaction, examine for blood and bile.

Blood (occult)—

In performing this test it is best to withdraw meat from the diet for 3 days beforehand, as muscle fibers give a positive reaction for blood.

Guaiac Test.—Take about 5 c.c. of stool made liquid with water unless already in a liquid state. Add 2 c.c. of glacial acetic acid. Shake thoroughly and let it stand for 5 minutes.

In another test tube take pinch of powdered gum guaiac and add to it 5 c.c. of 95 per cent alcohol, and let this stand 5 minutes. Filter. This makes tincture of guaiac. In the test tube containing the stools and acetic acid add equal amount of ether and invert tube carefully several times. Decant the ether. Add to the ether extract tincture of guaiac and hydrogen peroxide each about 3 c.c. Shake, and if you get a blue color it indicates the presence of blood.

Benzidine Test.—To a small amount of the fæces is added an equal amount of water. Take 3 or 4 c.c. of this and add to it 2 or 3 c.c. of alcoholic solution of benzidine (made up by saturating 95 per cent alcohol with benzidine, using some heat, then filter). Now add 1 or 2 c.c. of hydrogen peroxide and a few drops of acetic acid. In the presence of blood an intense green color develops. Occasionally a light blue tinge. This test is extremely delicate.

Bile (hydro-bilirubin)—

Make the stool liquid by addition of water unless already in a nearly liquid state. Take about 20 c.c. of this mixture and to it add 20 c.c. of a concentrated aqueous solution of corrosive sublimate. Shake well and set aside for 12 to 24 hours. Any fæcal particles that contain hydro-bilirubin are colored red, and all particles with bilirubin assume a greenish shade. This test is also of value when you wish to tell whether pale stools contain a large amount of fat, or is due to the absence of bile pigments.

Microscopic Examination

This includes food particles, epithelial cells, pus cells, blood corpuscles, bacteria, ova and larvæ of parasites.

The microscopic examination is rendered much easier if a portion of the fæces is made liquid by the addition of water, then filtered through 3 layers of gauze and centrifugalized for a minute. Pour off the water and add more and mix thoroughly with a glass rod. Centrifugalize for one-fourth minute

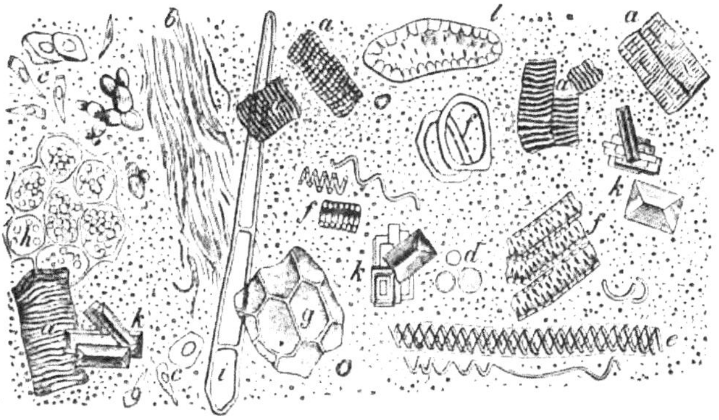

Fig. 22.—Microscopic elements of normal feces: *a*, Muscle-fibers; *b*, connective tissue; *c*, epithelial cells; *d*, white blood-corpuscles; *e*, spiral vessels of plants; *f-h*, vegetable cells; *i*, plant hairs; *k*, triple phosphate crystals; *l*, stone cells. Scattered among these elements are micro-organisms and débris (after v. Jaksch).

and again pour off the fluid. Repeat this a second time. Now with a pipette get the part from the bottom of the centrifugal tube, place this on a slide and cover with cover glass. This method is very useful when looking for ova of intestinal parasites.

1. Food Particles—

(a) *Muscle Fibers.*—Yellow in color, rounded ends, and usually in small particles only. No striations seen in well digested muscle fibers. In case of defective proteid digestion the striations are preserved.

(b) *Fibrous Connective Tissue.*—These are seen especially when there is a lack of free hydrochloric acid in the stomach —since they are digested by the gastric juice principally.

(c) *Vegetable Cells.*—Vary in shape, may be mistaken for intestinal parasites. The cells are more or less round or oval, less regular in shape, and not so uniform in size.

(d) *Vegetable Spirals.*—These are derived from certain plants which have escaped digestion in the intestines.

(e) *Vegetable Hairs.*—These resemble oftentimes embryos, but they differ in that they possess a homogenous wall, with a central canal extending throughout. They possess no motility —which living embryos do.

(f) *Starch.*—Occur as starch cells and starch granules. A starch cell possesses the characteristics of a vegetable cell described above. On the addition of Lugol's solution they appear blue. Starch granules are oval in shape, small in size, have a laminated appearance, and turn blue with Lugol's solution.

(g) *Debris.*—As a rule, considerable unrecognizable granular material is present—classified as debris.

2. Epithelial Cells—

These originate from the walls of the alimentary canal. They are of no importance beyond their recognition.

3. Pus Cells—

If intimately mixed with stools they are from high up the intestinal canal. If only observed microscopically they indicate a catarrhal or slight ulcerative condition.

4. Blood Corpuscles—

This is best recognized by the chemical test.

5. Bacteria—

The only one of practical importance is the bacillus tuberculosis. The method is as follows: Make a portion of the stool fluid with water. Take 12 c.c. of this and centrifugalize for 3 minutes. All the bacteria remain in the supernatant fluid. Pour off this fluid and add equal amount of 95 per cent alcohol, which changes the specific gravity to that lower than the bacteria. Centrifugalize a second time and this throws down the bacteria. Pour off the supernatant fluid and then make a thin smear of the sediment on a slide and stain in the usual method for the bacillus tuberculosis.

6. Ova of Intestinal Parasites—

It is very important to recognize ova of the various intestinal parasites. The ova most commonly found in the fæces are ova of the following: Oxyuris vermicularis, ascaris lumbricoides, trichocephalus dispar, uncinaria, tænia saginata, tænia nana, bothriocephalus latus, and tænia solium.

1. *Ova of Oxyuris Vermicularis* are often not found in the fæces. They are oval with rounded end and flattened on one side. In order to make a more accurate test for ova of this parasite, either scrape the mucous membrane of the rectum for material to examine microscopically, or give a cathartic and examine the fluid discharge.
2. *Ova of Ascaris Lumbricoides.*—They are oval in shape and brown or yellow from bile. The protoplasm is not segmented and is granular, and margin is uneven or wavy.
3. *Ova of Trichocephalus Dispar.*—They are oval, yellow-brown in color, and have a projection from each end which is the characteristic point.
4. *Ova of Uncinaria.*—They are oval in shape, not bile-stained, a transparent shell and a protoplasm usually divided into from 2 to 8 segments.
5. *Ova of Tania Saginata.*—They are oval in shape, contain hooklets in their protoplasm and a shell which is radially striated.
6. *Ova of Tania Nana.*—They are spherical, have a thick shell, which is usually in layers.
7. *Ova of Bothriocephalus Latus* (rare in this country).—They are oval in shape, brown in color, and have an operculum at one end.

Intestinal Parasites Recognized by the Parasite Itself, Portions of a Parasite, or Larvæ—

Intestinal Parasites that are recognized in the stool by the parasite itself, portions of a parasite or larvæ are: Amœba coli and cercomonas hominis (protozoa), segments of the different types of tape worms, ascaris lumbricoides, oxyuris vermicularis, uncinaria, larvæ of strongyloides intestinalis, larvæ of the house fly and trichina spiralis (rarely).

1. *Entamœba Histolytica* (protozoön).—Secure the grayish or bloodstreaked particles from a fresh stool passed into a warm vessel or by removal with a rectal tube. Place on a warm slide and immediately examine microscopically for their irregular amœboid movements.

2. *Cercomonas Hominis* (protozoön).—The adult organism is oval, and is provided anteriorly with a single long flagellum and posteriorly with a tail-like appendage. It is found associated with many active diarrheal conditions. Significance not determined. Probably not pathogenic. In fresh stools under microscope are actively motile.

3. *Trichomonas.*—The adult organism is oval or spindle-shaped. From its anterior end is given off four flagella. There is an undulating membrane extending down one side. The posterior pole is rounded or tapers to a tail-like appendage. Significance not determined. Probably not pathogenic. In fresh stools under microscope are actively motile.

4. *Segments of Tape Worms.*—Small segments are passed at intervals if an individual is infected with any of the types.

5. *Ascaris Lumbricoides.*—They are passed occasionally and as round worms are easily recognized.

6. *Oxyuris Vermicularis.*—Occasionally found in the fæces and about the anus of children.

7. *Uncinaria.*—Not often found in fæces. Give thymol, followed by magnesium sulphate, and they then may be usually found in an infected patient.

8. *Strongyloides Intestinalis.*—In this neither the parasite nor ova are found in the fæces, but the ova hatch in the intestines, and the actively motile larvæ may be seen. The stools should be examined microscopically while fresh and warm.

9. *Larvæ of the House Fly.*—The ova of the house fly may be ingested, and these hatch in the intestines. The larvæ, or maggots, may be found in the fæces as a result.

10. *Trichina Spiralis.*—The worm itself may be rarely found in the fæces, but the diagnosis is best made by the eosinophilia and the encysted larvæ in the muscles.

Fig. 23.—Ova of ascaris lumbricoides, with and without envelope.

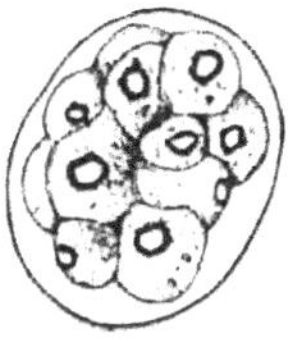

Fig. 24.—Ovum of uncinaria americana.

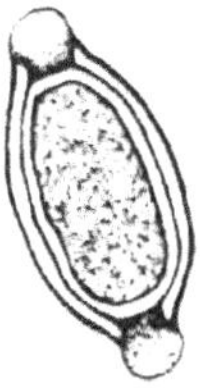

Fig. 25.—Ovum of trichocephalus dispar.

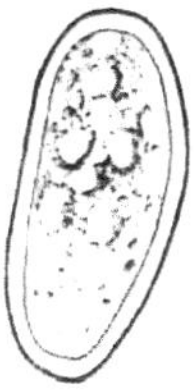

Fig. 26.—Ovum of oxyuris vermicularis.

Fig. 27.—Ovum of tænia saginata.

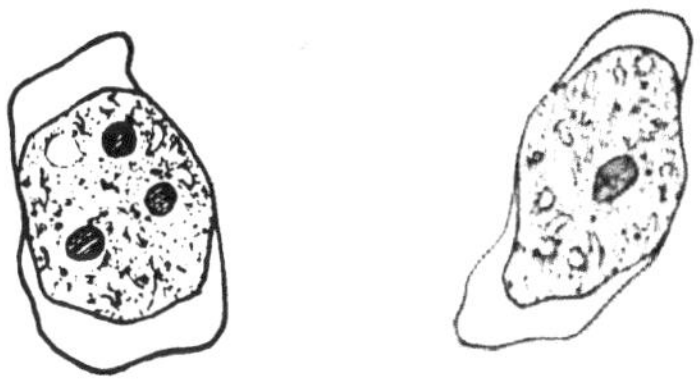

Fig. 28.—Entamœba histolytica.

Fig. 29.—Cercomonas homonis.

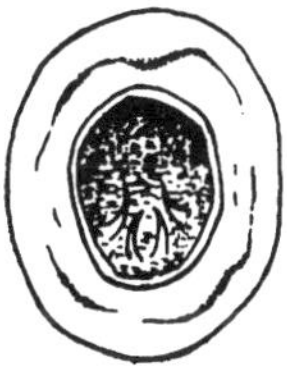

Fig. 30.—Ovum of tænia nana.

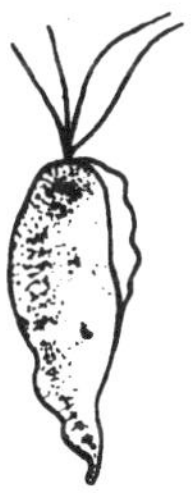

Fig. 31.—Trichomonas.

CHAPTER IX

TUBERCULIN DIAGNOSIS

This is employed in three ways:

1. Koch's subcutaneous method.
2. Cutaneous reaction of Von Pirquet, and Moro Ointment.
3. Ophthalmo reaction of Calmette.

Tuberculin is made by growing for 4 to 6 weeks a pure culture of bacillus tuberculous in 5 per cent glycerine bouillon. This is filtered and the filtrate evaporated to 1/10 of its volume. This resultant fluid is known as the tuberculin.

Koch's Subcutaneous Method

This is given subcutaneously in the back below the angle of the scapula. Ordinarily for the first injection 0.2 m. gm. is given; if no evidence of a reaction in 2 or 3 days a second injection of 1 m. gm. is given; in 2 or 3 days more 5 m. gms. are given, and finally 10 m. gms. Physiological salt solution is used as the diluting fluid.

The chief evidence of a positive reaction is fever, at least .5° C. or more. Other less important signs are: headache, malaise, insomnia, and nausea. A focal reaction may occur at the location of the tuberculous lesion. A local reaction may take place at the point of injection.

Indications for Use of This Method—

In adults with clinical symptoms, or clinically suspicious symptoms of tuberculosis, but who are devoid of the presence of tubercle bacilli and temperature.

Contra-Indications—

Fever, hæmoptysis, hæmaturia, marked cardiac or renal affection, arteriosclerosis, and diabetes. The patient should be placed in bed 2 or 3 days before the injection and temperature taken every 3 hours to be certain no fever is present. The patient should be kept in bed for 2 or 3 days following each injection.

Diagnostic Value—

A negative reaction following injection of 5 m. gms. is a very strong point against the presence of a tuberculous lesion in the body. A positive reaction indicates the presence of a tuberculous lesion, but whether it is an active progressive one or a latent lesion is difficult to say.

Von Pirquet Cutaneous Reaction

Cleanse the patient's forearm on the inner surface with ether. Place two drops of undiluted tuberculin upon the skin 10 c.m. apart. Scarify the skin first between the drops as a control, and then scarify within the two drops. Place on a piece of cotton for about 10 minutes.

Interpretation of Reaction—

Scarification of itself produces the so-called "traumatic reaction," i. e., a small wheal with rose colored margin. This passes away after several hours. The "specific reaction" is noticed upon the upper and lower points where the tuberculin has been applied and consists of a red indurated papule, which often extends in size 10 to 30 m.m. in diameter. This occurs within 24 hours.

Diagnostic Value—

In adults it is void of any diagnostic value, as 70 per cent of all adults give a positive reaction. Between the ages of 10 and 15 about 50 per cent of all cases react positively. It is of some value in children less than 10 years of age, and of very considerable value in children younger than 5 years.

Moro Ointment Reaction

A quantity of 50 per cent ointment of tuberculin about the size of a pea is warmed to 25° C., and this is rubbed into the skin of the abdomen for about one minute. A papule or small nodular eruption occurring within 24 or 36 hours is regarded as positive. The diagnostic value is variously interpreted, and it possesses in this respect about the same value as Von Pirquet's reaction.

Ophthalmo Reaction of Calmette

For this a one per cent fresh dilution of tuberculin in physiological salt solution is made. Place one drop of diluted tuberculin in the inner angle of the eye and let it run on the inner surface of the lower lid. In 24 hours, if mucosa of the lower lid is red and infected, the test is positive. This test is of particular value in all suspicious cases of tuberculosis where the presence of bacilli cannot be demonstrated, and the subcutaneous reaction cannot be undertaken on account of the presence of fever. It is used often in Vienna on ambulatory patients without any bad results whatsoever.

Contra-Indications—

It is contra-indicated in all diseases of the eye, tuberculous or otherwise. In case a second instillation is to be given the other eye should be used. This is sometimes done when there is no reaction with a one per cent solution. In this instance a two per cent or even three per cent may be used.

Diagnostic Value—

A positive reaction indicates in about 80 per cent of the cases the presence of tuberculosis, but as to whether the tubercular condition is active or latent it is impossible to say.

CHAPTER X

THE WASSERMANN REACTION

Preliminary preparation and tests for the Wassermann reaction are:

1. Preparation and standardization of the amboceptor.
2. Preparation and standardization of the antigen.
3. Obtaining and preparation of the complement.
4. Obtaining and preparation of control and suspected sera.
5. Preparation of diluting fluid.
6. Obtaining and washing sheep's red blood cells.

Preparation of Amboceptor—

Inject into the peritoneal cavity of a large rabbit gradually increasing quantities of washed sheep's red blood cells suspended in 0.8 per cent salt solution. Inject the first time 2 to 3 c.c. of the cells, in four days 3 to 4 c.c., in four days again 4 to 5 c.c., etc. About five or six injections are required. Five or six days after the last injection administer a little ether to the rabbit and dissect out the carotid artery, and collect the blood in two 50 c.c. graduates. Place on ice, and when the serum has separated, draw the serum off and place in 5 c.c. quantities in small size test tubes. The entire procedure must be carried out in a sterile manner. Place the test tubes containing the serum in a water bath at a temperature of 56 degrees Centigrade, and maintain this for 30 minutes. This inactivates the serum, i. e., destroys the complement.

Standardization of the Amboceptor—

Have six test tubes of about 10 c.c. capacity in a row, and number same reading from left to right as follows: I, II, III, IV, V, VI. Test tube I should have a capacity of at least 15 c.c. In test tube I place 0.1 c.c. of the amboceptor, then 0.9 c.c. of 0.8 per cent salt solution, and 9 c.c. of 0.8 per cent salt solution. This gives a dilution of 1 to 100. Carry 1 c.c. from this and place in test tube II, then add 1 c.c. of 0.8 per cent salt solution. Mix well. This gives a dilution of 1 to 200. Carry 1 c.c. from tube II to tube III and add 1 c.c. of 0.8 per cent salt solution. This gives a dilution of 1 to 400. Continue this process to tubes IV, V, and VI, which will give a dilution of 1 to 800, 1 to 1,600, and 1 to 3,200 respectively. Throw away 1 c.c. from the last tube. Now add 2 c.c. of 0.8 per cent salt solution to tubes II, III, IV, V, and VI; then add to same tubes 1 c.c. of complement, diluted 1 to 10 with 0.8 per cent salt solution. To the same tubes now add 1 c.c. of 5 per cent sheep's red blood cells in 0.8 per cent salt solution. Place in the incubator at 37½ degrees Centigrade for 30 minutes and note the extent of complete hæmolysis, using a dilution of the amboceptor between this dilution and the dilution in the previous tube. For example, if hæmolysis is complete in the 1 to 800 dilution and not present in the 1 to 1,600 dilution, then the strength to use is 1 to 600. This is the dilution to be used in the Wassermann test, and 1 c.c. is the quantity used.

Preparation of the Antigen—

Get a fresh heart from a healthy beef. Remove endocardium and epicardium. From heart muscle scrape off 30 gms. and place this in 270 gms. (by weight) of 95 per cent alcohol. Shake as often as possible for 24 hours and keep at room temperature. Then filter this through ordinary filter paper. This filtrate contains the antigen to be used in the test.

Standardization of the Antigen—

Have six test tubes of about 10 c.c. capacity, and number these I, II, III, IV, V, VI. In tube I place 0.1 c.c. of antigen, in tube II place 0.2 c.c. of antigen, in tube III place 0.3 c.c. of antigen, 0.4 c.c. in tube IV, 0.5 c.c. in tube V, and 0.6 c.c. in tube VI. Add to each tube 2 c.c. of 0.8 per cent salt solution. Then add 1 c.c. of the diluted amboceptor, usually 1 to 500, or 1 to 600 dilution, and next 1 c.c. of complement diluted 1 to 10 with 0.8 per cent solution, and finally to each tube 1 c.c. of 5 per cent sheep's red blood cells. Shake well. Place in incubator at 37½ degrees Centigrade for one-half hour. Examine to see what strength of antigen was inhibitive to hæmolysis. Usually 0.15 c.c. and 0.2 c.c. of antigen is used, as this is found practically always to be noninhibitory to hæmolysis.

Obtaining the Complement—

This is done by removing all the hairs from the under surface of the neck of a guinea-pig, and with scissors severing the vessels in the neck and catching the blood in a petri dish. A better method is to dissect out the carotid artery, and bleed into several test tubes. This blood is placed in ice chest, and the serum allowed to separate off. This serum is pipetted off and placed in test tubes. It may be necessary to centrifugalize to separate all of the serum.

Preparation of the Complement—

Dilute the complement in the proportion of 1 c.c. of guinea-pig's serum to 10 c.c. of 0.8 per cent salt solution. The complement should always be fresh.

Obtaining of Patient's Sera for Test—

Cleanse well the arm at the bend of the elbow, insert a needle into a prominent vein, and allow 4 or 5 c.c. of blood to pass into a test tube. Place in ice chest and allow the serum to separate. Place this at once into another test tube and inactivate by placing in a warm water bath at 56 degrees centigrade for 30 minutes. All this work should be done in a sterile manner.

Diluting Fluid—

This fluid is 0.8 per cent salt solution. Use C. P. sodium chloride dissolved in distilled water.

Obtaining and Washing Sheep Red Blood Cells—

Sheep red blood cells are obtained and washed in the following manner:

Clip wool from the neck of sheep in the region of jugular

vein. Shave this area. Place a cord about neck of sheep directly in front of front legs and tighten same so as to distend the veins. With a needle 5 or 6 centimeters long pass into vein and allow the blood (10 or 15 c.c.) to pass into a wide mouth bottle containing about 15 c.c. of 1 per cent sodium citrate solution in 0.8 per cent sodium chloride solution and a dozen glass pearls. Shake gently for several minutes. This is to prevent any clotting. Place in centrifugal tubes and centrifugalize about 8 minutes. With pipette draw off the supernatant salt and citrate solution. Pour in equal quantity of 0.8 per cent sodium chloride solution. Mix well with pipette. Repeat this process a second and a third time. The cells are now washed red blood cells. After the third washing add same amount 0.8 per cent salt solution as the red blood cells; this is the approximate quantity of blood serum. Two or three c.c. of this suspended in a little 0.8 per cent salt solution is used for the first injection into the rabbit. A 5 per cent solution is used in the standardization of the antigen and amboceptor, and also in the Wassermann test itself.

Technique of the Wassermann Test

Have a front, middle and rear test tube for each serum to be tested, also for positive control, negative control, and antigen control. In the front tube place 0.6 c.c. of the serum, then add to this 2.4 c.c. of 0.8 per cent sodium chloride solution. Mix well and then place 1 c.c. of this in the middle tube and also in the rear tube. This gives 0.2 c.c. of serum in each tube. Treat all the sera to be tested, and also positive and negative control, in this manner. In the antigen control tube place 1 c.c. of 0.8 per cent salt solution in front, middle and rear tubes.

In all of the tubes of the front row place 1 c.c. of 10 per cent complement, also 0.15 c.c. of antigen, 0.85 c.c. of 0.8 per cent salt solution.

In all of the tubes of the middle row place 1 c.c. of 10 per cent complement, also 0.2 c.c. of antigen, and .0.8 c.c. of 0.8 per cent salt solution.

In all of the tubes of the rear row place 1 c.c. of 10 per cent complement and 1 c.c. of 0.8 per cent salt solution, but no antigen.

Shake well. Then place all the tubes in the incubator at 37 degrees centigrade for one hour. In this time union between the antigen and syphilitic antibody (?) will have taken place in case the patient has syphilis, and so doing, fixes the complement.

In one hour this is removed from the incubator and 1 c.c. of the 1 to 500 (usually thereabouts) dilution of the amboceptor is added to every tube, and also 1 c.c. of the 5 per cent suspension of the red blood cells. The antigen quantity in the antigen control tubes is doubled. This is shaken well, returned to the incubator for no longer than 30 minutes, or until the controls come out properly.

Interpretation—

Hæmolysis should occur in the negative control tubes, also in the three antigen control tubes. The rear row of tubes should show hæmolysis. If any one of these fail to do so, the test on that serum should not be reported, but must be repeated. Positive control tubes should show the complete absence of hæmolysis. All the sera in which the two front tubes show the absence of hæmolysis are positive. All the sera showing complete hæmolysis are negative. The extent of the inhibition of hæmolysis varies from slight inhibition to complete inhibition. Therefore we say the Wassermann test is strongly positive when there is complete inhibition of hæmolysis—indicated by four plus (++++); is moderately positive when there is only slight hæmolysis, i. e., at least 75 per cent inhibition—indicated by three plus (+++); is weakly positive when there is 50 per cent inhibition of hæmolysis—indicated by two plus (++); is very weakly positive when there is 25 per cent inhibition of hæmolysis—indicated by a single plus (+); is of doubtful reaction of there is less than 25 per cent inhibition of hæmolysis —indicated by plus-minus (±).

The Wassermann test with cerebrospinal fluid is in every respect identical with that of blood serum, except it is not necessary to inactivate as it contains no complement. If blood (accidental) is present the fluid should be well centrifugalized, and if very much, also inactivated—1 to 2 c.c., or more, of the cerebrospinal fluid should be placed in each tube, instead of 0.2 c.c. as in case of the blood serum.

CHAPTER XI

COMPLEMENT FIXATION TEST FOR GONORRHEA

In this test the hæmolytic system is used as in the Wassermann reaction. The amboceptor is prepared and standardized exactly as in the Wassermann test. The complement is also obtained and prepared in the same manner, as is also the control and suspected sera. The sheep red blood cells are obtained and prepared exactly as in the Wassermann test. The antigen used by the author at the present time is that prepared by Parke, Davis & Company.

The test is as follows:

Have two rows of test tubes—a front row and a back row. In each tube of the front row place the following: 1 drop of undiluted antigen, 17 drops of 0.8 per cent salt solution, and 10 drops of complement diluted 1 to 10. In each tube of the rear row place the following: 18 drops of 0.8 per cent salt solution, 10 drops of complement diluted 1 to 10. Finally place in a front tube and the corresponding rear tube 2 drops of inactivated patient's serum. Repeat this in other tubes with inactivated positive serum, and also with inactivated negative serum.

This should be placed in the incubator at 37½ degrees centigrade for 30 minutes. Remove from incubator and place in each front tube 10 drops of the diluted amboceptor, the unit of which has been determined according to the method given in the Wassermann reaction, and 10 drops of 5 per cent washed sheep corpuscles. In each rear tube place also: 10 drops of diluted amboceptor and 10 drops of 5 per cent sheep corpuscles.

Place in the incubator at 37½ degrees centigrade and incubate until controls come out properly, i. e., all rear tubes should show complete hæmolysis. The front tube of the positive control should show the absence of hæmolysis. The front tube of the negative control should show hæmolysis. Therefore, any sera showing the lack of hæmolysis is regarded as positive, any showing the presence of hæmolysis is regarded as negative.

CHAPTER XII

TECHNIQUE OF STAINING AND EXAMINATION OF SMEARS, AND THE MOST IMPORTANT EXUDATES, ETC.

Gonorrheal Pus—

In every case of acute urethritis the discharge should be examined microscopically, since in a very small percentage of cases it is not due to the gonococcus. The gonococci are characteristically located within the cytoplasm of the pus cells and some in the epithelial cells, a few may be found free. The organisms occur in pairs and often several pairs are found in certain pus or epithelial cells. The Löffler's methylene blue and Gram's stain are employed for diagnosis. The latter especially as the gonococcus is gram negative.

Löffler's Methylene-Blue Stain—

Smear the pus on slide or cover slip.

Dry in air.

Fix by passing through flame 3 times.

Cover with Löffler's methylene-blue for 1 or 2 minutes.

Wash in water, blot, and dry.

If smear is on cover slip mount on slide with balsam.

Place on cedar oil.

Examine for intracellular diplococci.

Gram's Stain—

Cover the cover glass containing the smear with anilene water gentian violet for 2 or 3 minutes; blot dry.

Gram's iodine solution for 1½ minutes; blot dry.

95 per cent alcohol; pour on and off until all the blue color comes away.

Wash in water.

Cover with dilute carbol-fuchsin (1 to 2 with water) for ½ to 1 minute.

Wash in water, mount and examine. Here the diplococcus appears red from the contrast stain, dilute carbol-fuchsin.

Diphtheria Smear—

In obtaining material from throat be sure no antiseptic gargle has just been previously used. Let the swab come in good contact with the membrane or suspicious spot—avoiding touching other places as far as possible. Inoculate a Löffler's blood serum tube, and in addition make smear on a slide from the exudate or a piece of the membrane. It is possible to make immediate diagnosis from smear in about 25 per cent of the cases.

In case of smear, or in case of growth from the blood serum tube, two stains are valuable: Löffler's methylene-blue stain and Neisser's staining method. The latter is very satisfactory, especially in case of smears for immediate diagnosis.

Löffler's Methylene-Blue Stain for Diphtheria Bacilli—

Make smear as described above.

Dry in air.

Fix by passing through flame three times.

Cover with Löffler's methylene-blue for 2 minutes.

Wash, blot dry and examine.

With this stain the polar bodies at each end at times appear intense blue while the rest of the bacillus is a lighter color. In addition the beading is often clearly brought out.

Neisser's Stain for Diphtheria Bacilli—

Two solutions are required. Neisser's solution No. 1, and Neisser's solution No. 2. See Chapter XIII for composition.

Make smear as described above.

Fix by passing through flame 3 times.

Cover with Neisser's solution No. 1 for 30 to 60 seconds.

Wash.

Cover with Neisser's solution No. 2 for 30 seconds.

Wash in water, dry, and mount.

The bodies of the bacilli appear brown with dark blue dots at either end.

Exudates from Serous Cavities, (Pericardial, Pleural, Meningeal, Etc.)—

Make smear on cover slip or slide.

Dry in air.

Fix by passing through flame 3 times.

Stain with Löffler's methylene-blue, and also by Gram's method. See page 85 for exact technique.

Pericardial Exudate Secondary to Acute Inflammation—

Shows usually pneumococci, streptococci, or staphylococci.

Meningeal Exudate Secondary to an Acute Inflammation—

Shows most usually meningococci, or pneumococci. The former is a gram negative diplococcus found principally within the cytoplasm of leucocytes. The latter is a gram positive diplococcus, and found principally extracellular.

In tuberculous infections of any of the serous membranes one may, by high-speed centrifugalization, find the tubercle bacilli in sediment. Stain as for tubercle bacilli on page 21.

The most satisfactory and accurate method is to centrifugalize the fluid with high speed and inoculate subcutaneously with sediment in abdominal wall of guinea-pig. In 4 weeks guinea-pig will show signs of tuberculosis (enlargement of adjacent lymph nodes) and die within about 6 weeks in case the condition is of tuberculous origin. On autopsy the characteristic tubercles will be found in lymph nodes, spleen, and lungs.

Exudates from Furuncles, Abscesses, Carbuncles, Etc.—

Make smear on cover slip or slide.

Dry in air.

Fix by passing through flame 3 times.

Stain with Löffler's methylene-blue, and by Gram's stain. See page 85.

Most usually staphylococci and streptococci are found. These are gram positive.

Exudates from Eye—

Gonococci are responsible for the most serious acute inflammation of the eye. The organism is easily found in the smears from untreated cases. After treatment is begun, they very soon disappear. Make diagnosis by staining smear with Löffler's methylene-blue and by Gram's stain.

Koch-Weeks bacillus—a minute, slender bacillus—is responsible for acute infectious conjunctivitis. It is gram negative.

Staphylococci and streptococci most usually responsible for non-specific conjunctivitis.

Exudates from Ear—

In acute otitis media the organisms usually responsible are the pneumococci and the streptococci. The pus immediately following the incision of the drum should be smeared on cover glass and stained by Löffler's methylene-blue and by Gram's method. In this way the infecting organism will be easily, quickly and accurately determined. Discharges which have continued for some time are practically always contaminated with staphylococci.

Treponema Pallidum—

Giemsa's Stain.—Make smear as follows: Scrape crust off sore, if present, with small knife. Rub off with little cotton. Squeeze gently and obtain serum, with perhaps a very little blood. Draw this up in capillary pipette, and place one drop on a cover slip. Smear this over the cover slip. Make smear also on a slide in the same manner. Dry smear in air. Fix smear in absolute alcohol, 5 minutes. Remove and wash quickly in distilled water. Stain with dilute Giemsa's solution (alte Vorschrift-Gruebler)—i. e., 10 c.c. distilled water and 10 drops of the Giemsa solution—for 12 hours. Wash in distilled water. Blot dry with filter paper. Mount on slide, using liquid paraffin instead of balsam. Examine with oil immersion. The treponema appears blue with 8 to 10 uniform spirals.

India Ink Method for Treponema Pallidum—

Obtain serum from suspicious sore as described in preparation of smear for Giemsa's stain. Place one drop on a perfectly clean and polished glass slide. Add to this one drop of india ink (Gunther-Wagner black pearl ink) and then spread in a thin film by means of a second slide. Allow the slide to dry in air; it should be dark brown or black. Place on immersion oil and examine. The treponema is unstained and appears as a refractive spiral organism on the dark background.

The Luetin Test for Syphilis—

Luetin consists of cultures treponema pallidum sterilized, and then preserved with trikresol.

A very small drop of luetin is injected into the skin (not under it) of one arm. A like preparation without the treponema is injected into the skin of the other arm as a control. A positive reaction usually begins within 48 hours and consists of an inflammatory induration, papule, pustule. A positive test occurs, especially in tertiary, late secondary, and hereditary syphilis. It is inconstant in general paralysis and tabes dorsalis, absent in primary cases.

CHAPTER XIII

APPARATUS AND CHEMICAL REAGENTS NECESSARY FOR A PHYSICIAN'S LABORATORY

Apparatus

Microscope, with oil immersion lens, two-thirds and one-sixth objective; Centrifugal machine; Türck's hæmocytometer, with white and red cell pipettes; Tallquist's hæmoglobin scale; Cover glasses and slides; Burette; Cover glass forceps; Graduates; Specific gravity bulb; Doremus-Hinds urea apparatus; Test tubes; Red and blue litmus; Congo paper; Beakers; Flasks; Blood lancet; Filter stand; Esbach's tube; Funnels; Stomach tube; Pipettes; Water bath.

Chemical Reagents Needed

Concentrated nitric acid; Concentrated sulphuric acid; Concentrated hydrochloric acid; Glacial acetic acid; 50 per cent acetic acid; 25 per cent acetic acid; .5 per cent sodium nitrite; 10 per cent ammonium hydrate; Decinormal sodium hydrate; 10 per cent sodium hydrate; Sodium nitroprusside crystals; Aqueous solution of ferric chloride; Iodine solution (tincture iodine 1 part and alcohol 15 parts); 20 per cent lead acetate solution; Saturated sodium chloride solution; Saturated corrosive sublimate solution; Concentrated hydrochloric acid with .4 gm. of ferric chloride in 100 c.c.; Saturated aqueous solution of Bismarck brown; Wright's blood stain (0.3 gms. Wright's stain to 100 c.c. pure methyl alcohol). Absolute alcohol; 95 per cent alcohol; Ether; Chloroform; Löffler's methylene-blue; Gum guaiac; Tincture iodine; Hydrogen peroxide; Saturated aqueous solution of methylene-blue; Saturated alcoholic solution of acid fuchsin; Topfer's reagent (dimethyl-amido-ozo-benzol .5 per cent in 80 per cent alcohol); Gunzburg's reagent (phlorglucin 2 gms., vanillin 1 gm., 95 per cent alcohol 30 c.c.); Carbol-fuchsin (carbolic acid 5 c.c., saturated alcoholic solution of fuchsin 10 c.c., and distilled water 95 c.c.); 5 per cent copper sulphate solution; Strong sodium hydrate solution; 10 per cent copper sulphate solution; 5 per cent ammonium iron alum solution; 10 per cent ammonium hydrate solution; Saturated potassium oxalate solution; 10 per cent aqueous solution ferric chloride; Aniline oil; Saturated alcoholic solution of gentian violet; 5 per cent aqueous solution of carbolic acid; Crystals of sodium iodide; Infusorial earth; 4 per cent solution of glacial acetic acid tinged with little gentian violet; 10 per cent butyric acid solution in normal saline; Normal sodium hydroxide solution; 10 per cent potassium ferrocyanide; Oil of cedar for immersion lens; Xylol-balsam for mounting.

Fehling's Solution.—Dissolve 34.64 gms. of pure copper sulphate in water and make up to 500 c.c. Dissolve 173 gms. of Rochelle salts and 125 gms. of sodium hydrate each in 200 c.c. of water. Mix and make also to 500 c.c. Keep copper solution and alkaline solution in two separate bottles.

Sulphanilic Acid.—Make a saturated solution of sulphanilic acid in a 5 per cent hydrochloric acid solution. This is for the Diazo reaction.

Bromine Solution (For Urea).—Bromine 30 c.c., potassium bromide 30 gms., and distilled water 240 c.c.

Sodium Hydrate Solution (For Urea).—Sodium hydrate 10 gms., distilled water 250 c.c.

Hayem's Solution.—Mercuric chloride 0.5 gms., sodium chloride 1 gm., sodium sulphate 5 gms., distilled water 200 c.c.

Gram's Iodine Solution.—Iodine 1.0 gm., potassium iodide 2.0 gms., distilled water 300 c.c.

Esbach's Reagent.—Picric acid 1.0 gm., Citric acid 2 gms., water 100 c.c.

Phenylhydrazine Acetate Solution.—20 per cent aqueous solution of sodium acetate 25 c.c., 10 per cent aqueous solution of phenylhydrazine Hydrochloride 25 c.c. Mix.

Nylander's Reagent.—Bismuth subnitrate gms. 2, Rochelle salts 4 gms., 8 per cent sodium hydrate solution 100 c.c.

Standard Silver Nitrate Solution.—7.27 gms. to 250 c.c. of distilled water. 1 c.c. of this equals 10 M. gms. of sodium chloride.

Fresh Tincture of Guaiac.—Powdered gum guaiac in 95 per cent alcohol, shake well, then filter.

Decinormal Hydrochloric Acid Solution.—3 c.c. concentrated hydrochloric acid in 200 c.c. of distilled water.

Physiological Salt Solution.—0.85 gm. pure sodium chloride in 100 c.c. distilled water.

Standard Ammonium Sulphocyanate Solution.—3.25 gms. to 250 c.c. distilled water. 1 c.c. of this equals 1 c.c. of the silver nitrate solution.

Tsuchiya's Reagent.—Phosphotungstic acid 1.5 gm., concentrated hydrochloric acid 5.0 c.c., 95 per cent alcohol 95 c.c.

Acid Alcohol.—1½ c.c. of concentrated hydrochloric acid, 98½ c.c. of 95 per cent alcohol.

Rudisch Solution (For Quantitative Glucose).—200 c.c. of 50 per cent aqueous solution of potassium sulphocyanate, and 50 c.c. of a mixture of equal parts of Fehling's copper sulphate and alkaline solutions.

Neisser's Solution No. 1.—Methylene-blue 0.1 gm., alcohol 2 c.c., glacial acetic acid 5 c.c., distilled water 95 c.c. Dissolve the methylene-blue in the alcohol and add it to the acetic acid water mixture, then filter.

Neisser's Solution No. 2.—Bismarck-brown 0.2 gm., water (boiling) 100 c.c. Dissolve stain in boiling water and then filter.

INDEX

Spotted something that isn't quite right?

Let us know and we'll fix it for future reprints. And as a thank you for your diligence, we'll send you a free Sothis book of your choice.

publisher@sothis.press

This page intentionally left blank

Except of course for that line above. And this one.
And whilst we have your attention, perhaps you'd
like to follow us on Instagram @sothis.press. Go on.

Made in the USA
Las Vegas, NV
03 May 2026

46281660R30058